W9-DFF-108

Praise
for
Harvest the Rain

"*Harvest the Rain* is the book I have been waiting for: a detailed 'how to' for people and communities wanting to take a major step in saving the world's water written by a passionate water-conservation advocate. Let this practical, entertaining, and challenging book be your guide to your own—and the world's—water-secure future."

—Maude Barlow, author of *Blue Covenant* and senior adviser on water to the president of the United Nations General Assembly

"This book will not only make you a true believer in the regenerative power of harvesting rain—it will show you how. *Harvest the Rain* is full of practical solutions to our water shortages and points the way to a climate-resilient future. If we want thriving landscapes, abundant food, strong communities, and sustainable economies, we can start by treasuring rain."

—Andy Lipkis, founder and president of TreePeople and Ashoka Fellow

"*Harvest the Rain* is not just another true horror story about impending environmental apocalypse. Nor is it a dry how-to book about water use. Downey makes us feel the reverence that cultures throughout human history have felt for water falling from the sky."

—*Times Higher Education* (Tom Palaima, MacArthur Fellow, author, commentator, and professor of classics at the University of Texas, Austin)

"*Harvest the Rain* is a harmony of a passionate heart, a practical mind, and a spirit that communicates luminously."

—Richard Lederer, author of *Anguished English* and more than 30 other books about language

"The pleasures (and necessity) of cooking and eating are entirely dependent on water anywhere, but in dry regions like ours, the matter of water is an increasingly urgent concern. *Harvest the Rain* is an eloquent testimony to the necessity of taking local water seriously and learning to understand, harvest, and use it—important to read and more important to implement!"

—Deborah Madison, author of *Local Flavors: Cooking and Eating from America's Farmers' Markets*

"With an engaging voice, practical advice, and true depth of knowledge, Nate Downey offers real solutions to those who want to take control of their own water supply. *Harvest the Rain* should be required reading for anyone concerned about the future of water in this country."

—Carleen Madigan, author of *The Backyard Homestead*

"We as a species, of course, have to figure out how to live within our water means even as we transition from fossil fuels for our energy. *Harvest the Rain* is a great guide for individuals to learn how to succeed at that crucial task."

—Doug Fine, author of *Farewell, My Subaru*

"*Harvest the Rain* is a vital wake-up call to a healthier, more sustainable path available to us all. As Nate exclaims, 'Look! Up in the Sky! It's not a bird, a dog, or even a plane. It's rain!' Harvest it, as this book recommends, and precipitation becomes power."

—Brad Lancaster, author of *Rainwater Harvesting for Drylands and Beyond*

"*Harvest the Rain* is a new and optimistic look at our current issues concerning a vital resource—fresh water—and the urgent need to capture, conserve, recycle, and protect it in the face of the ferocious demands of this century. The book is practical, thoughtful, and passionate."

—Baker Morrow, FASLA, author of *A Dictionary of Landscape Architecture*

"Downey's anthem to the rain could do for the backyard and the water table—and therefore, let's hope, for the Earth and its inhabitants—what the *Joy of Cooking* did for the kitchen, or what *The Joy of Sex* did for the bedroom. It's one of those rare how-to books that, by way of the author's wit, warmth, and passion, converts practical wisdom into a kind of transformational incantation."

—Nick Paumgarten, staff writer, *The New Yorker*

Harvest the Rain

Harvest the Rain

How to Enrich Your Life
by Seeing Every Storm as a Resource

Nate Downey

Illustrated by George Lawrence

SUNSTONE PRESS

SANTA FE

Sunstone books may be purchased for educational, business, or sales promotional use.
For information please write: Special Markets Department, Sunstone Press,
P.O. Box 2321, Santa Fe, New Mexico 87504-2321.

Book and Cover design ♦Vicki Ahl
Body typeface ♦ CG Times and CG Omega
Printed on acid free paper

Library of Congress Cataloging-in-Publication Data

Downey, Nate, 1966-
  Harvest the rain : how to enrich your life by seeing every storm as a resource / by Nate
Downey ; illustrated by George Lawrence.
    p. cm.
  Includes index.
  ISBN 978-0-86534-495-2 (softcover : alk. paper)
  1. Water harvesting. 2. Rainwater. I. Title.
S619.W38D69 2010
628.1'1--dc22

                                          2010017330

Published in

WWW.SUNSTONEPRESS.COM
SUNSTONE PRESS / POST OFFICE BOX 2321 / SANTA FE, NM 87504-2321 /USA
(505) 988-4418 / ORDERS ONLY (800) 243-5644 / FAX (505) 988-1025

For
Melissa, my love,
from the bottom of my heart
to the top of the sky
in all directions
forever.

Contents

I sat upon the shore
Fishing, with the arid plain behind me
Shall I at least set my lands in order?
—T. S. Eliot, "The Waste Land"

Preface

Water as Stimulus

Don't forget your history.
Know your destiny:
In the abundance of water
The fool is thirsty.
—Bob Marley, "Rat Race"

When the ancient orchards are in bloom in the valley of Velarde and people are stepping out at first light to discover that their fragile buds have made it through the night, the high desert of northern New Mexico becomes one of the prettiest and most hopeful places on Earth. Here, the upper Espanola Valley bumps up abruptly against the bottom walls of the Taos Gorge as a snow-packed mountain backdrop puts life in perspective, slows thoughts down, and allows any soul the space to ponder one's role in the universe.

Half of my life I've lived in Santa Fe, the city of "holy faith," an hour's drive south. It's the oldest European-settled city in the United States, founded four centuries ago by *conquistadores* drawn toward fabled cities of gold and wealth. The Spanish found no gems or precious metals, but they stayed to till the verdant reaches of the Rio Grande highlands. Back then, tall grasses used to tickle the bellies of their horses, but due to many forms of cultural expansion and human "progress," few healthy grasslands still exist across the basin. The "big river" spreads out and floods naturally almost nowhere anymore. Progress, no longer synonymous with the proof of human greatness, is only a siren of change: sometimes for good, often for ill.

We watch our rivers diminish in volume while simultaneously increasing in levels of pollutants, turbidity, and sedimentation as they course from numerous headlands to the sea. Governments, farmers, ranchers, builders, environmentalists, utility companies, and all their lawyers fight over water rights that translate into drinking water, food, land, homes, endangered species, gas, oil, minerals, and jobs. Above all, we are witnessing how water will make and shape our futures, even as we watch overused water resources coming up short and water tables dropping across entire bioregions. Almost everywhere, this scenario replicates around the globe. Water, the stimulus of life, is the new gold.

Sometimes a side trip down Velarde's dusty roads reminds me why I'm so invested in issues of water, land, and sustainability. Such detours dare me to dream of a bountiful future when rivers and their peoples are respected. Other times I'm lured back to the fresh waters of my childhood when my grandmother used to take my older sister and me fishing down by the causeway at the bottom of Old Huckleberry Road at the southern edge of old rural Connecticut.

These good times still call out to me with pleasures children relish forever. After grabbing our rods from the garage and hitting up the compost pile for a cup of live worms, at daybreak Anne and I would shuffle off a few steps ahead, while Gramma stayed slightly back to keep an ear out for cars. On these short, steep treks she'd make us tote old shopping bags for collecting roadside trash. "A fisherman should always protect his watercourse," she'd proclaim as she yanked a dew-laden beer bottle from a gangly stand of grass.

Fat bass, muscular pike, and plenty of skinny sunnies devoured our bait. Now and then we'd save a chunky specimen for the evening meal, but almost everything we caught, we thought, was too much trouble to clean, so we'd grit our teeth and do our best to gently unhook the squirmy, supple, terrified vertebrates before quickly chucking them back into the dark water.

Later, up at the house, we'd squeal about those that got away and follow Gramma back to her piles of compost, where we'd help load her big red wheelbarrow. Then we'd drive it down the path to her dazzling flower garden. In front of an aged picket gate and simple lattice archway, Anne and I would swing, play, and giggle in the shade, while Gramma, trowel in hand, smiled in the sunshine.

At dinner, if we didn't eat fresh fish, I'd get Grampa to slice me a second "hunk of horse," as he once famously called Gramma's roast. With permission to be excused, we'd quickly scrape our scraps into a sink-side milk carton, rinse our plates, and dash off. After the grown-ups had their MacNeil/Lehrer fix, on the west-facing porch we'd drop gently into story time as the sun prepared to sink behind a precipitous ridge. As the last refractions of daylight dove out of sight, it was time to climb the steep stairs toward tooth brushing and bed. While she tucked us in, Gramma would whisper a simple prayer in French about the dreams of little angels, a mantra passed down to her by her father. Sometimes one of us would ask for a translation, but we didn't really need to hear the words in our native tongue. The deep love in our grandparents' voices transcended words and lulled us to sleep.

Fast forward a few years to an East Coast regionwide drought. I'm an eighth grader and our New York City sliver of an apartment is seven stories up, deep within a forest of steel, concrete, glass, and brick. Mom and Dad had recently surprised Anne and me by deciding to squeeze two more kids into our previously predictable family unit. With brother Jack now climbing the kitchen counters and soon-to-be sister Liz kicking around Mom's belly, I remember being relatively freaked out about the prospect that if it didn't start raining, we'd be part of a citywide, drought-induced exodus. I knew we were lucky to have our grandparents a quarter mile from freshwater, but I remember wondering: if those westerly storms kept dipping south and the rain kept missing us, what would our 20 million neighbors do?

Then one night on the phone I heard my grandparents say they could not recall a time when the reservoir had been lower. There would be no more fishing for a long time.

"Fishing," I gasped, "prohibited?"

When I later realized that almost all adults seemed disinterested in the water crisis, like any frustrated, self-respecting 13-year-old, I began to get genuinely righteous and sometimes livid as I mounted a water-conservation campaign.

Serendipitously, my homeroom teacher handed me the reins at our school newspaper, so I let fly an editorial tirade designed to shame people into changing their lifestyles.

"It will not kill you to cut down on water consumption," I raged in conclusion, "but it may kill you if you don't." My campaign didn't much succeed, but soon the weather changed back to soggy and the habits of the tri-state area reverted to its typical state of relative nonchalance toward water issues.

Today I find myself having lived in a very arid land for over two decades. Here, the reservoirs are off limits. Forget fishing—you can't even get past the fencing and warning signs that surround Santa Fe's surface-water supply. During the spring snowmelt and sometimes during our short summer monsoon season, the usually brown, sediment-filled Santa Fe River rages wildly, but for at least 300 days per year the river's watercourse is a glorified trickle that meanders gradually out of town, looking more like a road than a river, toward the Rio Grande. I don't get angry anymore at people who lack my anxiety about water, but I often think of an old and true line about how it is always time to try to make a difference even if the big picture seems beyond any of us: "It is the most ridiculous of all mistakes to do nothing because you think you can only do a little."

Maybe it's the thin air I've been breathing at 7,000 feet above sea level, or it could be the high traces of lithium rumored to pervade parts of our local aquifers, but my decision to do what I can do on the water front, while maintaining a relatively calm composure with respect to a growing water crisis, could simply be a basic survival strategy at work. Over my long career as a water-conscious landscape designer, I have learned that I couldn't force change on anyone—especially anyone who might be writing me a check. As a conversation topic, "our role in the environment" is always fair game during any landscape consultation, but it's hard to know exactly when another person will be open to seeing their property in the context of an intricate web of concepts like "a healthy water cycle," "seven generations in the future," "aquifer independence," and "your local foodshed."

Even though my company, Santa Fe Permaculture, attracts a certain mindful clientele, not everyone is always ready to dive into the world of precipitation collection, conveyance,

"It is the most ridiculous of all mistakes to do nothing because you think you can only do a little."

storage, and distribution—much less to delve deeply into the metaphysics of sustainability or a hands-on biology lesson provided by a little bucket full of kitchen scraps. Lately, however, people seem more than willing to get deeply green in personal, meaningful ways that improve their lives, their homes, their neighborhoods, and the well-being of our planet. Somehow inspired by the current economic meltdown, more of our clients are unabashedly articulating a brand-new commitment to environmentalism. They understand that shifts in their individual perspectives and personal lifestyles are required, and they are expressing a much deeper desire to make the necessary changes. Landowners at every scale are seeing their property more as an investment they *must* nurture, overcoming financial hardship with a focus that begins at home and points to the future. Within the confines of having, as always, limited time and, more recently, a lower-than-expected budget, many of my clients are now seeing gardening, landscape design, and water harvesting as opportunities to embrace rather than as options to avoid.

Although the economy, global warming, healthcare, illiteracy, the flu, and a number of other serious challenges are important topics for discourse, exposition, alarm, and even outrage, this book mostly avoids fear and loathing as tools for change. My sense is that enough of us are coming to understand our predicament. A tipping point is near at hand. We just need conduits for animating our innate desire to thrive. When I think about all of the energy that can be put toward people who want to be part of "the solution," I realize there isn't time to go backward to hammer out blame, anger, fear, or righteousness. Instead it's time to get to work using the language of *convenience, return on investment, empowerment,* and *pleasure.* The ground *has* shifted.

While our society gains a greater reverence for water, it will be necessary to simultaneously develop deep esteem for all things sustainable: everything from local food and energy production to appropriate transportation and green building. There are countless causes to choose from, but water issues are especially exciting because anyone can relate to the nurturing power of water, and each of us can easily have a positive effect on our regional watersheds. As a consequence, everything from our local economies to the health of the planet as a whole becomes better. After oxygen, water (in most climates and during most times of year) is our prime need. Even before food and shelter, water is at the innermost core of any successful group of human beings.

The majority of this book describes a variety of ways to harvest rain—as rainwater, snowmelt, hailstones, sleet pellets, dewdrops, fog particles, and every other form of precipitation imaginable. But before we consider how to capture, save, and use this supply in the sky, I will first describe some of the incentives that come with collecting, storing, and distributing the most powerful and most easily accessible resource around.

1. **Water harvesting offers** *convenience*. By using a simple scheduling system that I call "gradual greening," harvesting large quantities of water becomes a convenient way to do one's part to save the world while at the same time getting a little exercise, making friends, improving one's quality of life, and/or investing in oneself and one's property. Imagine a world where we spend, on average, over four hours per day making our lives and our communities more sustainable. Now imagine such a world where no one really recognizes this time as anything special because it seems like second nature to us. This is all possible if we commit, on average, 10 minutes per day to gradual greening and if we renew this commitment every year (with an additional 10 minutes per day) for the next 30 years.

2. **Water harvesting provides** *return on investment*. By using my "get rich slowly" plan, you can turn any property into a relatively high-yielding financial asset over the long term. Beauty, shade, privacy, wind protection, and food production— these are just a few of the many ways you can improve the value of your home while also benefiting your local community and every living creature on this planet. With a modicum of patience, trees, gardens, outdoor rooms, and other landscape features will take root, grow, bloom, and mature. Although these improvements are often low cost to install, they provide high-yielding returns.

3. **Water harvesting** *empowers* **people.** Water and water harvesting in particular represent vital issues around

which people can organize. By producing water, preventing pollution, and creating green-collar jobs for plumbers, architects, landscape professionals, educators, manufacturers, arborists, scientists, laborers, and many more, water harvesting can be a populist, pro-environment and pro-growth plank to any political platform. It's hard to imagine an issue on which ultra conservatives, adamant moderates, and radical liberals can so easily agree. Not since the Great Depression has there been such a desire to shift course, to work together, and to use government to benefit societal goals. Seemingly insoluble problems are giving way to a horizon of great opportunity. Now is the time to get empowered in our communities and articulate new visions of water's universal relevance in a sustainable future.

4. **Water harvesting translates into** *enjoyment*. Harvesting precipitation is a pleasurable experience, from straight-on *fun* to the measured satisfaction of "game changing" accomplishment. From mulching your backyard with your family on a Saturday afternoon to working with friends and neighbors on a seasonal watershed-restoration project, water-saving projects act to facilitate a deep, enjoyable love for life. The creativity involved in landscape design, the physical exercise associated with gardening, the camaraderie generated among folks who come together to plant trees or to learn a useful technique—these are just a few of the countless examples of joy that come with water harvesting. Of the many roads to happiness, there are few as swift and dependable as the path through a healthy garden. In such a place, especially one that you had a hand in creating, delight, the mother of all incentives, morphs into clear reality. Is there anything much more fulfilling than the feeling we experience whenever we make an attempt to improve our world for future generations?

Introduction

Look! Up in the Sky!

Let it be said by our children's children that when we were tested we refused to let this journey end, that we did not turn back nor did we falter; and with eyes fixed on the horizon and God's grace upon us, we carried forth that great gift of freedom and delivered it safely to future generations.

—President Barack Obama in his inaugural address, January 20, 2009

Considering the worldwide strife caused by battles over fossil fuels, struggles for clean water and food have been given little press. Yet water and food, as much if not more than oil, will be the limiting resources for humanity in the 21st century. In terms of the sustainability of modern human culture, water is the most serious issue of our time. British author and journalist Fred Pearce in his powerful book *When the Rivers Run Dry* describes how people often go to war and civilizations frequently fall when water supplies fail. If history is any guide, unless we reverse the current water arithmetic, our modern culture will eventually break down. Aquifers, rivers, lakes, and reservoirs are shrinking rapidly, and many of our once-ample water supplies are being mined, polluted, and/or sold off to multinational corporations motivated by maximizing profits, not the common good. In addition to wreaking havoc on water-stressed areas, global warming, multiple security issues, armed conflicts, geopolitical shifts, and massive human migrations will also place increased pressure on the resources of "water rich" nations, regions, and communities.

Fortunately, there is a convenient, enjoyable, empowering, and potentially profitable approach to combat this situation. In your own backyard and within your own local community, you can collect, convey, store, treat, and distribute precipitation in a wide variety of ways. Wherever you are, you can harvest the rain and become an honest and productive steward of the Earth just like so many generations have before us.

Water harvesting is typically divided into two categories: passive water harvesting and active water harvesting. The term *passive* refers to the lack of moving parts in the system. This basic and essential form of water harvesting holds moisture in the soil such that anyone with a patch of land can easily start harvesting precipitation. Chapter 2, "Passive Water Harvesting Techniques," looks at a very wide range of techniques from landscape design and mulching to building swales and creating moisture-conscious microclimates.

Active water harvesting systems divert precipitation to storage tanks called "cisterns." The term *active* refers to the moving parts that are often required for removing water from storage. Anyone who would like to take their passive water harvesting yield to a higher level of precision and productivity will find chapter 3, "Active Water Harvesting Systems," particularly helpful.

In the long run, if we wish to create and maintain sustainable lifestyles during the thirsty times ahead, my proposed water harvesting revolution must go far beyond passive and active harvesting. It must also embrace on-site endeavors to reuse wastewater as it simultaneously supports community-wide efforts to make local watersheds regenerative. These two future-oriented approaches are described in chapter 4, "Wastewater Harvesting Methods," and chapter 5, "Community Water Harvesting Opportunities."

My firm belief is that it will probably be society's sewage-treatment technicians, our valiant plumbers, who finish the job of saving the world that passive, active, and community water harvesters have already started all over the planet. Right now, relatively simple wastewater treatment systems can be installed on a variety of scales at reasonable costs, but few people know about them. These systems are especially important because they have the potential to close the water-use loop for a household, which can ultimately translate into aquifer independence, or water self-sufficiency.

Community water harvesting is a term I'm coining to describe a more subtle form of water harvesting. Chapter 5 lays out a smorgasbord of ways to harvest precipitation on a scale that's bigger than any normal person's backyard. From schools that teach organic farming and beekeeping to political campaigns that change the face of nations and neighborhoods, this kind of water harvesting provides an option for those of us who like to work in groups (and who may or may not feel drawn to a designer's pencil, a gardener's rake, or a plumber's wrench). As the human paradigm begins to shift toward an enlightened respect for water, community work will become critically important in not only getting the necessary word out but also in ensuring that the requisite work, at every scale, is accomplished.

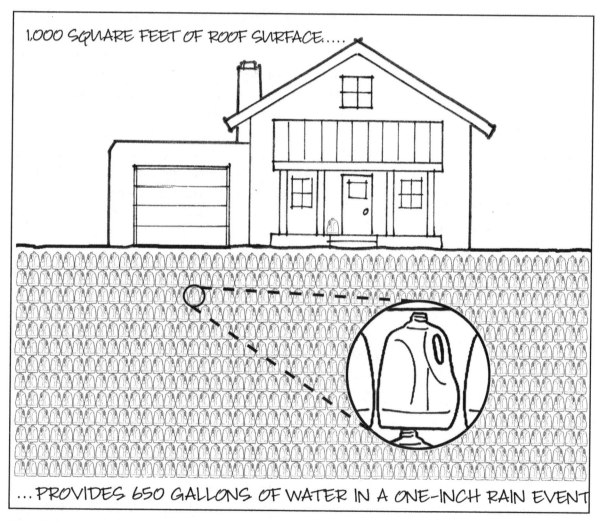

Rain Power

The seriousness of our worldwide water woes plus the inherent incapacities of desalination, conservation, privatization, and modern science all add up to the fact that we need a new approach to life's essential liquid. We need a fresh perception of precipitation. Like the Copernican revolution that changed the way many people perceived reality, the imminent water harvesting revolution is as radical but nowhere near as polarizing. Simply put, instead of looking downward for all of our water, we must now also look skyward, where anyone with a 1,000-square-foot patch of land has the potential to harvest thousands of gallons of extremely fresh water annually.

Other than the fact that many smart and sane people believe there is a water crisis looming for our culture, why should anyone make precipitation, or water in general, such a high priority? Why should we care about snow, sleet, hail, fog, dew, and the thousand names for rain? Why bother in the first place? For if there is no personal motivation to harvest rain, many people will simply choose to do other things.

First, on an ethical level, we should bother because what good are all of our carbon-neutral buildings, biofuels, eco-schools, and green-collar jobs if our freshwater supplies run dry? Water is essential for our survival as a species, so ensuring that it be provided to future generations is a moral imperative. This means we had better get working on a permanent solution to our water problem if we want to look our children directly in their eyes ever again.

Water is essential for our survival as a species, so ensuring that it be provided to future generations is a moral imperative.

Second, speaking practically, I like to eat, and I know farmers can't grow crops or tend livestock without plenty of water. For this reason, this book concerns itself with more than merely elevating our consciousness with respect to our household water. In addition to encouraging watershed protection, job creation, eco-education, and the creation of a more valuable, beautiful, and comfortable physical environment, this book stresses the importance of water harvesting in order to support local food production.

Third, from an ecological perch, harvested water requires far fewer emissions per gallon of water produced when compared to pumping an aquifer, river, lake, or reservoir. The imbedded energy of these conventional water supplies translates into extremely significant percentages of the human carbon footprint, so precipitation collection helps prevent global warming by reducing our dependence on the energy used to lift and deliver water.

Meanwhile, the biodiversity of any place is often negatively affected when we mine and divert water, the life blood of nature. And, as more species die as a direct result of their source of life drying up, the existence of those of us higher up on the food chain becomes more tenuous.

Equally important may be the fact that the massive, man-made water-storage systems that we call reservoirs contribute to global warming in a significant manner. By releasing astounding quantities of greenhouse gases in the forms of methane (CH_4), carbon dioxide (CO_2), and nitrous oxide (NO_2), the largest dam financier on the planet, the World Bank, reports that large-scale hydroelectricity has an adverse impact on our planet's temperature and the consequential effects of most big reservoirs should be regarded as similar to a typical coal-burning power plant. The problem is that flora and fauna, particularly in the large quantities often found at subalpine elevations, are not supposed to experience death by drowning. Unfortunately, as biomass decays in water, significant quantities of methane are produced, and as emissions go, methane is 20 times more problematic as a greenhouse gas than carbon dioxide.

Fourth, from a self-interested perspective, precipitation is the purest beverage in the world. As long as air-quality standards are maintained, clean collection surfaces are used, and a relatively simple filtration system is installed, rainwater is almost always safer and healthier than groundwater or surface water. Additionally, precipitation is one of the few salt-free sources of water available, and this makes it a boon to backyard gardeners and farmers alike.

Fifth, from a communitarian standpoint, water harvesting is a very diverse subject that provides numerous avenues for participation and a *common ground for conversation*. This community-creating quality that water harvesting has may be the field's most endearing aspect. Like few other topics of discussion, water, especially in times of perceived need, creates a deep bond among people. Even if enough of us were to simply talk about rainwater harvesting, future generations would build on our dreams and improve upon our successes. Understood in this manner, passing on information about water harvesting becomes among the most important roles of any sustainable society. Starting with something as simple and fundamental as water harvesting, it suddenly seems possible to leave a world behind for our children that is better than the one we received or at least much less prone to drying up and blowing away—like so many societies that have popped up and disappeared before our time. So the questions simply become: Will we talk about rain *enough*? Will an adequate number of "civilized" human beings know what to do when water problems get worse? And will we, now, build a vernacular around water harvesting that leads us toward a sustainable, comfortable, and pleasurable existence?

Sixth, within the context of other proposed solutions, I turn to water harvesting for the answer in part because the alternatives to precipitation collection and reuse have a much lower probability for providing a workable solution. One sometimes-trumpeted panacea, the desalination of seawater (and other brackish resources), will not be a viable long-term option in most parts of the world for a very, very long time if ever. The removal of salt from saltwater is an economic possibility only in low-lying coastal countries possessing very large quantities of cheap energy. Oil-rich Saudi Arabia, the prime model of this approach, desalinates 25 percent of the annual seawater that currently gets separated worldwide. Compared to the bounty that we can harvest from our atmosphere, desalination is too expensive to take seriously as a widely utilized fix. At the same time, unlike water harvesting, desalination causes pollution, does nothing to improve the value of adjacent land, and brings no additional beauty, comfort, and happiness to our communities. Desalination may generate big profits for large corporations that want people to be dependent on them, but in most places, it will simply not be a realistic option. Such facilities are also subject to devastating terrorist attacks and therefore also require expensive security operations. Water harvesting is completely localized, so terrorists would be much less able to damage the water supplies associated with the pending water harvesting revolution that I envision.

Water conservation, another "solution" to the problem, is essential, but it should also be recognized as a subtle form of procrastination that puts the onus of finding an actual way out on some wiser, thirstier generation. In fact the whole notion of conserving water might be more effective if we called it waste prevention. "Conservation" makes it sound as if we are doing the world a favor by conserving instead of honestly admitting the sad fact that we waste far too much water in our daily lives.

The ultimate solution to ensuring human life in the long term will revolve around ways of *collecting* and *producing* water locally, not merely *conserving* it. In most watersheds, especially those with growing human communities, "conservation" means simply reducing our resources at a slower pace rather than maintaining or increasing our supply. I do not wish to minimize the importance of conservation—it not only buys us the crucial time we need to set up sustainable societies, but it also underscores the reverence that we must achieve with respect to the lubricant of life.

I only wish to make the point that conservation is not enough. When it comes to our water resources, we must also begin to become responsible for our own backyards, neighborhoods, and watersheds. As my friend Pamela Mang, cofounder of the Santa Fe–based planning firm Regenesis Group, made it plain to a handful of guests at a green-business forum at her home, "We must become much more than mere preservationists because it's not enough just to conserve a particular resource. The human race has got to become a regenerative species again. Indigenous peoples have always understood the positive, productive, and contributing role that human beings can play in natural systems. We have to get back to that or the systems that we depend on for our existence will continue to degenerate, and many will soon fail completely."

Mang even goes so far as to say that the conservation mind-set can actually be a detrimental notion if it separates humans from nature in an unnatural way. "The conservation movement has tended to think of nature as some *other* thing that needs protection, but this disconnection is an illusion that ignores the interconnectedness of all living things." Although I am not necessarily ready to say that conservation is part of the problem, the critical point is that we cannot preserve our way to sustainability. Human activities, such as the collection, storage, and reuse of precipitation, that jumpstart nature's ability to regenerate itself are certainly far more essential than actions like water conservation that merely postpone the inevitable.

Water privatization, the most ludicrous suggestion to have been proposed as an answer to the water question, is sometimes advertised as a cure-all for the worldwide water

The ultimate solution to ensuring human life in the long term will revolve around ways of collecting *and* producing *water locally, not merely* conserving *it.*

crisis, but in fact it is a poorly publicized swindle that powerful corporations are trying to pull around the globe. Motivated by corporate charters that demand short-term profit, companies like Suez and Veolia sell water to the highest bidder. With immense financial backing from the World Bank, the World Trade Organization, the U.S. Agency for International Development, and many other global organizations (including the United Nations), the trend until only very recently has been "Locals be damned. Let's make money!"

Couched as being efficient, conserving, and generous, the existing small international water cartel is exactly the wrong kind of organism for the best management of the shrinking water supplies found in every country across the globe. As Canadian author and activist Maude Barlow makes clear in *Blue Covenant*, "water warriors" throughout the world are standing up to these powerful interests. A founding mother of a growing water-justice movement, Barlow, along with her allies, is slowly easing the grip of this cartel. Having recently served as the senior adviser on water to the president of the United Nations General Assembly, she believes a better way to handle the world's water would be to give local water sources to the people who walk, work, and fish up, down, and all around their local watersheds. This seems like a reasonable suggestion, does it not?

Many people believe that technological advancements hold the key to our water future. Although there are promising high-tech approaches that could reduce our need to harvest rain, most are at the margins of practicality. In a few cases, I have incorporated them into my broad definition of water harvesting. For example, atmospheric water generators that harvest moisture from air are discussed later in chapter 2 in "Harvest the Air."

Some of the best scientific research related to water harvesting is being done by Dr. Andrew Parker and his colleagues at Oxford University's Green College. A leading proponent of biomimetics (engineering that mimics nature), Parker's work copying the fog-collection strategies of dung beetles and the moisture-sucking abilities of some lizard species will no doubt make water harvesting more efficient and productive. Certainly, scientific discovery should be part of any water harvester's toolbox, but most of the technology put forth in this book is of the low-tech variety because *Harvest the Rain* is designed for everyone everywhere. Even forgetting the possibilities that industry might someday provide, there are enough low-tech solutions to our water problems that already exist. If we would only use them, we would not have to wait with hope against hope for salvation from the same technology that has gotten us into this mess. Advanced technology is and will always be useful, but to depend on it completely is not only unnecessary, but it's also foolish.

Hovering on a more mystical plane, my seventh reason for my wanting people to begin to revere and respect precipitation is that water should be seen as a sacred element from the infinite reaches of space to the core of any being's consciousness. In addition to the practical, ethical, economical, social, and political arguments for diverting our attention toward precipitation, there is a spiritual component at work that is difficult to deny. Water, particularly precipitation in the desert, seems to sanctify the soul in an equal proportion to the way in which it benefits any living body.

Many indigenous tribes still perform daily religious ablutions in nearby rivers just before dawn. Serious readers of the Qur'an require a watery *wudu,* or cleansing, before glancing at the good book's pages. From the holy water of a Christian baptism to the spiritual cleansing associated with a wade through the Ganges River, seas get parted en route to promised lands, rain clouds get hammered by gods of thunder, and bowls of water overflow at funeral services of Buddhist monks. It's almost as if to be a religion, some strong connection to water is required or people won't even bother. We can even add powerful secular belief systems to this list as faithful capitalists revere liquid assets, pious politicians bow to watershed moments, and ancient medicine and modern healthcare both prescribe H_2O at every opportunity. Looking at water in this way, one could say that water moves the human soul as much if not more than any other substance.

If you are unsure where to find the humid center of your particular form of faith, do not hesitate to ask your neighborhood sage for some water-based guidance. He or she will probably be happy to oblige.

Happen to be between religious communities? Don't have a guru to turn to? Are you and your spiritual teachers at a loss for water-based sacraments? If so, please feel free to try the following exercise. Even if you are already completely comfortable with your relationship to water, you might enjoy it. The exercise doesn't take long, and if it doesn't irritate or confuse you too much, it may even prepare you appropriately for the leisurely but powerful approach to water harvesting suggested in chapter 1.

Please.
Stop reading this,
and have a glass of water.
Feel it soak into your being. There.
You have just empowered the life force.
You have just reconfirmed your will to live
and have acted on behalf of the creator
of nature and consciousness. Water
is spirit. Respect, even worship,
water in its purest form.
Precipitation is
power.

Designed to elevate our appreciation for rain, this exercise intends to drive home our dependence on water as it simultaneously attempts to make us recognize the awesome act of imbibing. We take our origins for granted and do not recognize their central significance in our physical, economic, social, political, and spiritual lives. Further, the necessary shift in our cultural paradigm depends at least in part on an increased level of respect for the chief sponsor of natural growth, that abundance in the heavens. Three-hundred and sixty-five days a year, a certain potential in the sky waits for a few million particles of dust, a simple change in temperature, and the chance to dive back down to become an essential part of some life-form again. My own belief is that the mere thought of water, especially as it cycles through time, is enough to significantly rev the human soul and give it the metaphysical juice needed for amped-up spiritual growth, powerful creativity, and a profound form of almost überhuman compassion.

The comic-book creators of cultural hits like *Superman* and *Underdog* are right when they suggest that we "Look! Up in the sky!" for heroes that might save society from imminent doom.

They are wrong, however, in proclaiming that this assistance will come in the form of an extraterrestrial bodybuilder in a red, gold, and blue unitard or in the agreeable guise of an articulate, pill-popping beagle. It's much simpler than that. All we need is a down-to-earth change of perception, a simple switch of perspective, a watershed moment of the mind: one that provides us with a clear realization as to what water really is and what it must begin to mean to us as we become a more sustainable species.

Precipitation is the source of life and consciousness. We just need to realize this fact. Look! Up in the sky! It's not a bird, or a dog, or even a plane. It's rain!

1

What the Power of Precipitation Can Do for You

How poor are they who have not patience!
What wound did ever heal but by degrees?
—William Shakespeare, *Othello*

You Can Grow Green

Every plan to save civilization from itself needs to provide a vehicle for transporting people from point A, a low place of frustration, to point B, a high perch of knowledge and understanding. Since water enables growth and lifts spirits up, in *Harvest the Rain* we will look at a mode of transport and a path to wisdom that are grounded in water harvesting. It's a system I call "gradual greening."

I start by asking you for a personal commitment—to donate 10 minutes of your time each day to water harvesting. Allowing for one skipped day per week, this will equal one hour of time donated for every seven days of your life. Assuming you take a yearly two-week vacation, your annual donation of time will equal about 50 hours per year. The central idea of gradual greening is to provide a bite-size challenge that also happens to be a profound opportunity to save the world, one day at a time: If you *add another 10 minutes to your daily routine every year and continue adding 10 minutes every year* (more or less—as best you can given your circumstances), at the end of 20 or 30 years you will be harvesting precipitation on average about three to four hours per day—less than the amount of time that we as a people currently watch television and about as much as many commuters spend in the Friday-afternoon rush home.

Three or four hours every day may seem impossible at first but if you gradually add time to your days for healthy outdoor activities (and nearly all forms of water harvesting deliver great exercise for the heart), you'll come to what will seem like a perfectly fair amount of time dedicated to the survival of your community, our species, and nature as we know her. Fortunately, with so many incentives associated with gradual greening, this slow and steady transition toward sustainability is doable. Just keep in mind that your main goal should be to slightly increase your level of dedication each year.

Harvest Your Brain

I provide many "tips of the trade" for this gradual greening—diverse techniques, systems, methods, and opportunities described throughout this book. You set the clock and you run the show. Gradual greening is a way to see the world and a way of *becoming*. Greening provides a wide variety of avenues of effectiveness from which people can choose, so that there is something for everybody at every stage of their development.

For example, you don't have to be a gardener, plumber, architect, or home builder to start harvesting rain in your community. If you simply donate some time every month to a water-wise nonprofit organization, you just joined the club. Personally, I lean toward local groups. From my perspective here in northern New Mexico, this suggests groups like River Source, WildEarth Guardians, or Camino de Paz School and Farm, but it could also mean getting involved in locally based groups with global appeal such as the Bioneers and Earth Care International. If you do not already have your own personal favorite local group that you'd like to support, then *today* would be a great day to discover some of the organizations making a difference on the "water front" of your neighborhood, community, or bioregion.

Some may say that gradual greening takes too long. Many will say that we have to dedicate much more of our time right now "before it's too late." The problem is, until many more people face the consequences of inaction, it will be left to each of us to make a difference as best we can. Often a nearly magnetic attraction to modern conveniences will trump our ecological concerns and our motivation to be green. We tend to get stuck in our ways, a thermodynamic law of motion controlling our souls, operating against our own best interest on any rational plane of thought. Aristotle warned 2,300 years ago that habits are difficult, almost impossible, to change. We may say we want change, but unless forced by events or coerced, we are by nature slow on the uptake.

Writing a generation earlier, Plato, in his *Republic,* redirects his classic conversation about the "perfect city" toward the critical question concerning "creature comforts." When Socrates discovers that his interlocutors are unwilling to give up the enjoyable, but energy-intensive aspects of conscious existence, he shows us how human nature's powerful desire for the relishes of life interfere with the successful pursuit of pure virtue.

Both of these ancient philosophers hit the mark. Under normal circumstances, human beings are not going to change quickly, and most will be entirely unwillingly to give up cherished customs overnight for some glorified ditch-digger's "water wise" paradigm. In the big picture, change takes time, but each of us can change and gradually green ourselves, our homes, our communities, and our world. We just have to employ more patience and forgiveness than we are accustomed to, and as a reward for keeping better tabs on our own precipitation-conscious actions, for starters we should probably stop, breathe, and sip more water more often.

As far as the 10 minutes a day goes, understand that this is just an average. We must start today with whatever time we can spare. The important point isn't *how* we divide up this small time commitment. It's that we increase our 10-minute-or-so commitment each year over the course of our lives. Sure, if we gently encourage others to do the same, so much the better, but let's face it: to expect big, immediate, and voluntary change is naive.

Some people will make water harvesting a big part of their livelihoods. Others will have time for a water harvesting hobby, and still others won't have much time at all, but each of us has the potential to change, to raise our consciousness ever so slightly, and to recognize the value of our experience. Everyone has the power to profit from any investment of time. Depending on your lifestyle, this might translate into spreading your commitment more or less evenly over the course of a week, or it might mean you condense your time into larger chunks. For some people, it might make sense to set plans a year in advance in order to have one work-week's worth of time off when you do a backyard makeover, to install or supervise the installation of a cistern system, or to tackle a greywater recycling project.

In the end, it doesn't really matter what shade of green you become or what shape your water harvesting endeavors take, nor does it matter *when* during each year you do your greening. What matters is *that* you begin and then remain *steadfast in your commitment*. Finally, keep in mind that any time you attempt to learn any form of "ecology," you are putting in time as a "gradual greenie." You are more than an immobile tree hugger. You are blessed with positive energy. You're a work in progress; you're a productive conservationist not only conserving but also growing as a person, creating resources, rejuvenating a local watershed, and making the world a better and more productive place. Every season of every year you're a living, breathing green machine: wise, water respecting, and walking the "beauty way," as the Diné, the Navajo people who have survived desert life for thousands of years, called the path of wisdom and awareness.

You Can Get Rich Slowly

At either end of the reaction spectrum, there are two basic ways to address problems as they arise: you can either grieve about your new situation or you can focus on all of the opportunities that exist in the midst of changing circumstances. My "get rich slowly" plan is a simple scheme for reacting to economic challenges in a productive manner. It centers on the fact that you can vastly improve the value of any piece of real estate as long as you have patience and some decent incentives.

No matter where you find yourself at this historical moment, let's assume that your particular property has a lawn. Per person we spend about $700 annually on the American lawn. I propose that families shoot for spending this lawn budget on any form of water harvesting that I describe throughout this book. If you haven't been spending money on your lawn, consider diverting what you spend on cable television, DVD (rentals and purchases), video games, pay-per-view entertainment, and the gamut of electronic components associated with screen-oriented amusement. In 2003 the figure was close to $100 billion—$300 for every man, woman, and child in America.

Want to watch your money multiply in myriad ways? Then exterminate your television and your lawn mower, and every year you will have, on average for every person in your household, $1,000 extra to spend on your property. Instead of wasting your hard-earned cash on "lawn inputs" (such as water, gas-powered mowing, poisonous chemicals, and expensive labor) or the idiot box (such as a new flat screen, the latest cell phone, that upgraded laptop, or next-generation video game) why not consider harvesting rain on your property in the wide variety of productive ways that I describe below?

Cultivating plant material with the water you harvest

If you happen to be new to gardening, start with a trip to every plant nursery in your area. Think like a "natural investor."

is a particularly good example of a sound investment in your property. Plants are inexpensive, require relatively little maintenance, and provide amazing returns. With beneficial trees, shrubs, vegetables, herbs, and flowers that are also well placed, you'll get incredible curb appeal, impressive erosion control, comfortable shade, effective wind protection, important privacy, constant noise abatement, beauty, fragrance, food, firewood, wildlife habitat, and a lower water bill than the typical lawn. And with any of these effects, you'll be increasing the value of your home, especially if you decide to take a mid- or long-term outlook on your property.

If you happen to be new to gardening, start with a trip to every plant nursery in your area. Think like a "natural investor." You'll find that nurseries are hotbeds of ideas that will enrich your property over time. Think short and long term. Garden experts are usually eager to share their experience and knowledge. Focus your questions on what plants thrive in which microclimates in your region. Listen carefully, but don't expect to absorb a local nurseryperson's wisdom without taking notes, if not also pictures, and even, ironically, using a video camera.

If you happen not to be motivated by yard work, it's also important to realize that the water harvesting industry will continue to grow in the foreseeable future. More and more firms that help people harvest their own precipitation will provide "investment services" wherever water resources are limited. In a down economy, however, every project will be highly sought after, which can translate into better-than-usual service from hungry contractors, who will have no choice but to impress their clients if they plan to survive. In a challenging economy, this is a plus for any nongardener who recognizes the need to invest in one's land in the short term in order to receive the greatest possible return in the long term.

The current upheavals in the U.S. and international markets, beginning with mortgages and equity financing, are doing more than threatening business as usual. Many are now talking about a new type of capitalism emerging from the worldwide meltdown of market economies. It is time, without doubt, for fundamental change. New approaches and new solutions are possible in ways that were not possible a brief time ago.

This book's approach is grassroots. Although the

problems economically and ecologically are immense, so are the opportunities. In the future, increasing percentages of homeowners, local lenders, real estate developers, and planning commissioners will understand the importance not just of "curb appeal" but also of a concept I call "blurb appeal." Ten to 30 years from now my guess is that the property with the successfully applied get-rich-slowly plan will find buyers before the property with the lousy, 20th-century lawn. With all of its traditional curb appeal, turf grass symbolizes wasted resources and pollution, while a property with blurb appeal might boast, "Beautiful shade trees, prolific berry bushes, gravity-fed cistern, and low-maintenance greywater recycling system provides privacy screening as it creates two intimate outdoor living spaces and fantastic curb appeal." With all of this going for it, which property will be more desirable down the road? The one that features an expensive and expansive lawn or the one that provides the real appeal of sustainability—perhaps in a short blurb next to a pretty picture of a cozy home tucked behind a veritable Eden.

One lending institution based in my local community already understands the inherent value of water harvesting, solar energy production, and the lowered maintenance costs of sustainable systems. It's the Permaculture Credit Union (PCU) based in Santa Fe, New Mexico. Able to lend anywhere within the United States and its territories, the PCU gets a little more ink in the community water harvesting part of this book (chapter 5), but it's worth mentioning here at the outset how this pioneering institution has had (for many years) a "sustainability discount," which reduces its standard loan rates by a certain percent for projects that can demonstrate sustainable objectives.

"It's a great way to promote sustainability," says Permaculture Credit Union president and CEO Donald J. Sarich, "because it makes people realize that some projects put money back into their own pockets. Passive solar architecture reduces heating costs. Hybrid cars use less gas. Cisterns save money on water bills. It all makes sense if you're like most of our members, who all have houses, cars, and at least a little bit of landscaping to take care of."

Looking back on my tenure as chairman of the board of the PCU, my best moment was hiring Sarich. He's been the credit union's president and CEO since 2003 and has been extremely successful at building an institution that is both sustainable and at the cutting edge of the eco-lending world. Perhaps the best part is that, as a credit union, money in the PCU is controlled by a democratically elected board of directors focused on long-term growth and financial viability, not by stockholders looking for short-term gain.

In addition to making money on real estate, in the vast majority of business environments there is also tremendous potential for successful businesses that recognize the productive power of precipitation. In addition to being a book for every homeowner, *Harvest the Rain* is for entrepreneurs, too—particularly for precipitation-oriented companies in a weak housing market. Most of the book's short chapters describe a water-based practice that can be perfected by any niche-oriented businessperson, so I encourage readers to go beyond their own homes and consider the opportunities that exist in their local communities. There is profit potential associated with green goods and services; you just have to discover what the particular potential is in your neck of the woods.

Many years before he became President Barak Obama's chief of staff, Rahm Emanuel said, "You don't ever want a crisis to go to waste." From the perspective of these interesting times, the message here might be that we can choose to let economic times affect us adversely or we can decide, as best we can, to take advantage of bad times and shift the power of our every investment toward alternative energy, local food production, and water harvesting on a human scale. It is not hyperbole to say a truly green revolution is possible in this lifetime. Whatever the politics of big government and corpulent corporate dinosaurs, it makes sense to divert a significant portion of our personal incomes to activities that will enhance the life within our specific household and increase the value of whatever patch of real estate for which we happen to be responsible. This is especially easy to do when we see precipitation as the resource that it is.

You Can Become Empowered

Just as I suggest a way to gradually green your time on this planet, and just as I promote a deliberate way to invest in your land and its associated structures, there is a slow and steady way to reinvigorate society and stabilize local economies. Almost everywhere, taxpayers spend an extraordinary amount of money on water projects funded at all levels of government. This money is often squandered on ventures that are not sustainable in the long term. Every day, dams silt up, canals fill with salt, rivers disappear, reservoirs get polluted, and wells run dry.

Even though gravity often plays an essential role in the movement of water through our infrastructure, typically in New Mexico one in four barrels of oil consumed is used to move water. In addition, consider the annual expenditures that your city, county, state, and federal governments currently allocate toward the design, construction, and maintenance of unsustainable water systems. Projects like Santa Fe's San Juan Diversion Project, at a projected cost of $425 million (not including the interest on borrowed money), is being built to divert a portion of a tributary of the Colorado River to households and businesses in Santa Fe. Via a tunnel underneath the Continental Divide the water already travels to the upper reaches of the Chama River, which later pours into the Rio Grande. The city's goal is merely to get its "rightful" Colorado water out of the Rio Grande and pump it uphill to its rate-paying and tax-paying constituents. If we transferred a thirtieth of this $425 million every year, we could instead invest $14,000 into cistern systems for 1,000 households in the city per year. By doing this, in 30 years Santa Fe could become dependent on underground and surface-water supplies only in times of severe drought.

When it comes to redirecting public monies, the key to true community empowerment will again be to move at a slow and steady pace. First of all, we need to envision the longer term and create local success stories as we overcome regulatory hurdles, finagle through governmental bureaucracies, and testify as to the virtues of budgetary allocations for renewable water projects. We begin with the current water infrastructure, wherever we are, and an understanding that we do not want to completely abandon our connections to surface and groundwater supplies. We will always need these in times of drought and other expediencies, but we also want the option of using, as much as practical, the easily accessible water supply from the sky.

A boom in water harvesting technologies and installation companies will of course bring its own challenges. For example, the reputation of the solar-water-heating industry in New Mexico was devastated in the late 1970s and early 1980s when installers began selling systems for tax-deduction purposes instead of for energy-savings purposes. Enticing tax credits pushed spending for poorly engineered and constructed home systems before enough people could get trained in solar technology and installation. When systems soon stopped working, as they regularly and quickly began to fail, and when roofs and walls started to leak due to the panels that had been improperly placed on buildings, an entire industry was tarnished for many years. Most of the solar-water-heating and -energy companies disappeared forever.

A scenario of trial-and-error and off-and-on failures will likely happen in a variety of places during the impending water harvesting revolution. Often out of necessity, local industries will sprout up more quickly than the workforce and permitting agencies can be brought up to speed. Sometimes water harvesting systems will get installed improperly, tanks will leak, foundations of buildings will sag, lawsuits will come down, and the water harvesting industry's reputation may even become tarnished at a local level. Fortunately, since most water harvesting projects (especially for landscaping purposes) do not involve complicated science and advanced engineering, the overall reputation of water harvesting will prove over time to be about projects that are built properly, that meet the homeowners' realistic goals, that increase the value of their associated properties, and that get replicated across the nation and around the world. But this will depend on contractors, developers, and their friends in power avoiding the lure of a quick buck and focusing on long-term success and further-down-the-road profit.

Each local community will be different. Needs, desires for growth, available financing, skilled entrepreneurs, community leaders, activists, and homeowners all become important factors at play in any community's ability to change. Some communities will need vocational education services in water harvesting training at various levels. Community colleges will offer extension courses. Other communities will initiate seed funding for pilot projects on the ground. Some communities will develop a powerful local water harvesting industry, standing out in their state and/or bioregion and encouraging the further growth of a support network around ecological projects and new urbanist developments. Outmoded building codes will be challenged and visionary political leaders with green supporters will set a new agenda for property development and community planning.

Although most of the techniques described in *Harvest the Rain* require only a nominal investment, some of the larger systems require significant cash. The water harvesting constituency will have to learn to be careful not to demand too much money too quickly from governmental sources. Appropriate growth plans will aim to set achievable goals. Political strategies will revolve around identifying line items in government budgets that can be scaled up to provide recurring revenues and extended direct and indirect government support.

One problem is that conventional water systems represent tremendous pet projects for elected officials. These are the kinds of projects that are often dangled at voters as jobs programs at election time, and they are the kinds of ventures that often translate into political war-chest donations. Unions are not removed from this equation either. Trade unions regularly are beneficiaries of bridge-to-nowhere-style pork-barrel projects that have been known to taint the reputations of politicos and big businessmen alike. But *blue-green alliances* are springing up everywhere with win-win solutions that provide ecological jobs of value to any community.

As Maude Barlow in *Blue Covenant* describes, there are thousands of water warriors in the water-justice movement who are already on the ground making a difference in their water-stressed communities. The time is now for water harvesters to ally with them. When we do, this will change the political terrain and our natural landscapes simultaneously for the better. It doesn't matter if you ever touch a shovel, a pipe, or a rain barrel. Especially in these days of online computer tools, forums, electronic social networks, and high-speed interactive communications, if you can organize people politically around sustainable water systems, your work will become an empowering political force in your community.

Since no community can exist without a water supply and a local, even personal (rooftop) water supply is the ultimate

Water harvesting's upsides are multiple and impossible to ignore. Harvesting produces water resources, prevents pollution, and creates a constellation of jobs.

in "planning for a rainy day," water will always be among the most important issues in any community. Water is motivational since clean water is an essential matter of survival. Health and water go together, and water often determines when a community is "first world" or "third world." Water issues "may be out of sight, out of mind," but take away clean water and suddenly its absence makes the gut grow fonder, particularly as people get desperate.

Water harvesting's upsides are multiple and impossible to ignore. Harvesting produces water resources, prevents pollution, and creates a constellation of jobs in value-added niches. Politically, water's a winner. Water brings people together, everyone from folks concerned about their jobs to people worried about endangered species.

Community organizing around water supply, water quality, and water sustainability will be central in the human communities of the latter part of this century, so water activism must now become a perennial cause wherever supplies are stressed. In this way, as issues come up, the grassroots can be contacted and political pressure can be exerted in the form of e-mails, letters, phone calls, meetings, legislative lobbying, public testimony, and, if necessary, steps-of-the-capitol demonstrations. Citizens volunteering, running for elections, writing, and putting forward platforms all have an essential place in the empowered communities of our sustainable future.

You Can Enjoy Life

As we begin to understand the ways in which precipitation is an essential resource, we begin to become part of the water harvesting solution. If you see the potential for health, wealth, and power that water harvesting can bring, consider yourself a water harvester and an important cog in the waterwheel of our survival as a reasonably civilized species.

Although the concepts of gradual greening, getting rich slowly, and empowering your neighbors in a deliberate way are helpful, like the strategies that we are about to delve into, they are, by no means, *required* concepts for your particular version of successful sustainability. Your ideas and specific projects have their own unique potential, their own power to produce water and water-related benefits. Your individual green dreams and water-conscious plans are connected to a larger community at work, increasing the overall level of water availability, productivity, and biodiversity throughout your bioregion and beyond.

Almost by accident, the work we do in the garden or for our larger watersheds can increase the amount of water that ends up in a local aquifer while simultaneously improving the quality of the water that reaches our streams, rivers, ponds, lakes, and reservoirs. By preventing pollutants from careening into our water supplies and by allowing some of the precipitation that we harvest to percolate into underground reserves, water harvesting can have a profound effect. Conserving for today and for future generations is work that connects us with those souls who will come after us and who will undoubtedly appreciate our work.

Like surgeons stitching up a wounded body, water harvesters heal the land. This seems like a powerful cosmic gift and, as it happens, I tend to believe that something at the essence of the universe is somehow grateful when we clearly understand water's essential role in all that lives and in consciousness itself. Water is sublime. Call it the "essence" of life, God, Allah, Yahweh, Buddha, Brahman, Zeus, Odin, karma, nature, creator, prime mover, or the original cosmological cause that provided a natural cradle, nurturing and making possible life forms on our planet, an oasis in a vast universe. A universal spirit, assuming such a thing exists, must (wouldn't it seem right and just?) appreciate those who do their best to keep life alive.

If you start collecting, storing, and redistributing water as "rain from the heavens," that is, as a powerful life force, then it's not much of a leap in belief to rethink this spiritual aspect beyond water's many material benefits. For many of the same reasons that your property appreciates in value when it has a beautiful, functioning, and efficient landscape, your quality of life tends to increase accordingly. From the positive effects of shade trees and view screens to the precious time you gain gardening with family and friends, the pleasures to be found in your water-borne garden are replete and life affirming.

Every new collection method opens up a conversation as to how to improve our water harvesting skills, how to design a more water-productive home, or how to plan a more water-sustainable community. When children grow up seeing sustainability at work, the pace of change will begin to grow exponentially. Generation to generation the lessons of nurtured life are passed on, and a socioecological legacy will be created as it continues to build momentum toward success.

Success is in the trees we plant, the gardens we grow, the water harvesting systems we build, the green communities we create. Everything we do to improve our immediate natural environment has the potential to bring a lyrical peace, a surprising joy, and a positive purpose to our existence.

Step into this garden of delight. You can do it. You can design and improve your environment in ways that calm the soul, spark the senses, and spread the spirit and knowledge we need in order to sustain our culture in the future.

High in the clouds, like an angel, a wellspring of joy, the fountainhead of our sustainability, the secret of our long-term human survival is waiting to be fully appreciated, duly honored, and thoroughly enjoyed. Some say she goes by the name of *rain*. Others say, "Water." Clearly a child of Earth and Sun, she lingers as long as possible with Mother Earth until summoned up to a swirling atmosphere somewhere between Venus and Mars. After immaculate acts of evaporation, condensation, and precipitation, the cycle continues, and it grows life. One day some descendents of all of this wonderful moisture discover consciousness and a way to pass on more of this life-giving liquid to as many generations as the heavens might allow. With this knowledge, life becomes a true pleasure once again for everyone who partakes in the power of precipitation.

2

Passive Water Harvesting Techniques

Before enlightenment, chop wood and carry water;
After enlightenment, chop wood and carry water.
—Zen koan

Start Small

It was a cold and freakishly foggy November morning. Bill Mollison, the most well-known progenitor of a school of thought (and action) called "permaculture," had begun a daylong seminar focusing, ostensibly, on a regional problem of rapid and pervasive evergreen tree die-off. Over the previous two and a half years across an area covering much of northern New Mexico and three neighboring states, 90 percent of the piñon tree population had been eviscerated by a sap-hungry beetle. Mollison, Australia's official "Ecologist of the 20th Century," had been invited to shed light on the drastic situation.

After an initial description of cosines, moon cycles, and the predictability of drought patterns, the white-bearded anti-guru, without warning, bombarded his audience with a verbal campaign of shock and awe. Still grieving over the death of so many neighborhood trees, many attendees were deeply offended by Mollison's tirade on a host of seemingly tangential topics: our lazy citizenry, the world's "fairy worshipping" religions, the inherent evil of industrialized tofu, and the "cancer" of the American lawn. Some students squirmed when Mollison segued to the issue of parents who purchase ponies as a covert form of birth control for their preteen daughters. Others became visibly nauseous when they were told about how the dog food they buy often comes from cows that have been slaughtered on the other side of a fence from some of those small groups of starving children featured in ads sponsored by Save the Children.

Just before lunch, having provided his desired context, Mollison began discussing "the departed." Surrounded by an endless army of zombie pines flaunting sharp ochre needles and pathetic pallid boughs, he clasped his hands below his tummy, stuck out his wide, wiggling thumbs, and beamed broadly.

"I think," he smiled and paused, "it's marvelous."

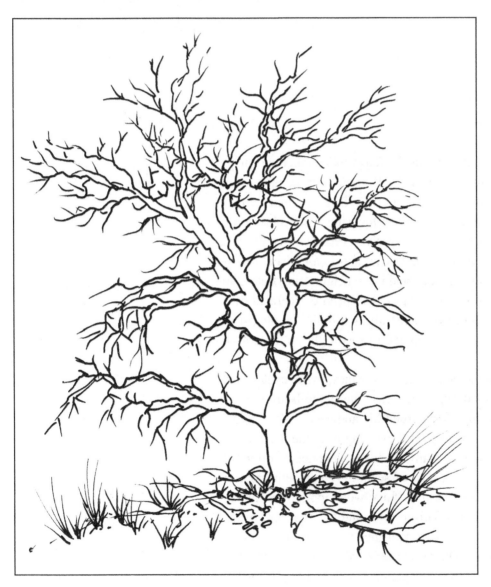

Marvelous in Death

Although I'm not entirely sure what caused it, the shift that occurred within me at that moment was monumental. It may have been the rapid movement of Mollison's brow between his seemingly serious "I think" and his dazed but gleeful "it's marvelous." Or it could have been Mollison's loud and congested laughter that echoed over the moaning groans of his increasingly disheartened audience. But suddenly I was beginning to understand something that Mollison had attempted to teach me years earlier: start small.

Most seminar attendees expected less callousness from a guy who had been so instrumental in building the beginnings of a worldwide sustainability movement. Some probably needed a hint of sympathy and respectful understanding in the face of so much localized forest destruction. But Mollison proudly admits he is not a gentleman. "I've seen too much devastation and ecological breakdown, far worse than your little tree tragedy," he told his anguished audience.

Over half of those in attendance ducked out during lunch, and a steady stream stole away before day's end; one had to wonder if Mollison could have benefited from a few manners. Still, the lessons the hangers-on learned in the final moments of the presentation made the experience worth it. "Every *drop* of water that flows from a piece of land represents a failure in our design," Mollison concluded.

That's when it clicked for me. For over a decade I'd been an ecological landscaper in the arid mountains and dry mesas of New Mexico. As a frequent harvester of rain and snow, it was only then that I realized the full potential of water harvesting. The moral of the Mollison message is not just about collecting and distributing water from storm events; it's also about the everyday, almost invisible times when dew, fog, condensation, relative humidity, shade, and wind come into play. Most importantly, it's about designing landscapes with the goal of harvesting every drop of moisture available.

Passive water harvesting holds moisture where it is needed most—near the root systems of beneficial plants. This retention simultaneously reduces sedimentary pollution in our surface-water supplies and can even replenish local aquifers. Successful passive water harvesting springs from the installation of a series of intelligently placed materials that are typically small, low-cost measures that reap impressive rewards.

Design with Nature

When it comes to the art of storing water in the soil, perfection is a long way off for any of us, but this is most especially true when planning (budgeting, scheduling, and design) have been neglected. All of the techniques and relatively simple systems explained in *Harvest the Rain* will never reach their true potential without a detailed and comprehensive landscape plan. Bill Mollison's *Permaculture: A Designers' Manual* is one of the best places to start for details in this regard.

Mollison says, and he's right, that you can design any system to be more efficient and productive if you apply ecological principles and natural patterns to your own *four-dimensional* planning. Whether designing a sustainable business, bureaucracy, household, community, political campaign, garden, farm, or wildlife habitat, the bedrock of permacultural thought is that the designs of our systems should always mimic nature or else resources will be wasted and potential benefits will never be actualized.

Mollison delves deeply into about a dozen ecological principles and as many powerful natural patterns in his manual. Here I'll focus on two that have obvious connections to water harvesting.

> Principle 1. Every component of a design should function in multiple ways.
>
> Principle 2. Every essential function should be supported by many of the system's components.

The first principle refers to what is often called Mollison's "synergy" principle, while the second principle suggests the well-known concept of "redundancy."

Synergy, which occurs whenever a whole is greater than the sum of its individual parts, is found everywhere in passive water harvesting. One can think of passive water harvesting integrated with erosion control, watershed purification, aquifer regeneration, localized biodiversification, and carbon-footprint reduction. Each method is integral to a precipitation-collection system functioning at an optimal level.

This kind of synergy is compounded whenever people

apply numerous passive water harvesting techniques in a holistic system. Whenever good soil, rich compost, thick mulch, deep swales, properly placed gabions, native seeding, deep shade, and effective windbreaks are combined, the quantity of harvested precipitation is always far greater than it would be when compared to the moisture generated by any individual water harvesting component.

A successful water harvesting plan must also include redundancy. You never know when one of your water sources might not be available. Passive water harvesting provides many additional links in a chain that will continue to function even when the system is weakened by drought, pollution, leakage, overconsumption, regulation, ordinance, or law.

The basic idea is that when one of your property's essential functions, for example, providing water, is only supported by one component, such as a well, this is a mistake. What if the well were to run dry or become contaminated? Or what if its pump needs to be replaced or necessary parts are days away? On-site wastewater harvesting, an active water harvesting system, some pumice wicks, a series of swales, and some healthy soil blanketed by a thick mulch—these all create a web of redundancy that makes sustainability possible during hard times. Similarly, a home could have passive solar glazing and plenty of thermal mass, but it wouldn't hurt to have a back-up woodstove for those occasional extra-long strings of cloudy days. Candles, plenty of thick blankets, and a couple of warm bodies never hurt either.

Almost by definition, synergy and redundancy will be designed into every efficient and productive landscape. Look for them also in every successful workplace from restaurants, convenience stores, and dental offices to bureaucracies, big boxes, and factory assembly lines. You can even take these principles and apply them to a kitchen or closet renovation with tremendous success, but there may be no better place to use these concepts than on a piece of property for the purpose of absorbing moisture.

Observe with Every Nerve

During times of abundant rainfall, it's easy for people to overlook the benefits of rain harvesting. Rivers flow high and fast, reservoirs fill quickly, and the earth around us greens up gorgeously. Amidst the surge of friendly smiles, especially in dry climes, it's as if people are being invigorated with every passing cloud.

The reality, however, is that in most cases we are diminishing and/or polluting our water supplies much faster than nature can replenish and cleanse them. While always fully appreciating every moment of good fortune and great abundance that we are fortunate enough to get from the sky, the ethical imperative with regard to future generations is clear. We should be looking at, listening to, and reaching for ways to harvest precipitation during floods, droughts, and every kind of weather in between.

Wet times, however, should not only be appreciated for the water they bring, they also provide a welcome deluge of information as to where water collects and moves on your property. During the next big rain in your area, be sure to get out and walk around your house. Take off your shoes. Wiggle your toes in the wet earth. Slide down a slope. Roll around. Lavish your body with mud. Taste the wet grass. Inhale the smell of fresh moss. Close your eyes and listen to the calm happiness permeating your rejuvenated neighborhood. Hug a big tree. Absorb the cool breeze. Savor the sight of that high cloud on the horizon.

Where is it warm? Where is it windy? Where is it wet? Observe all of the places where water flows on your property.

Just as no one entered Plato's school of philosophy without first observing and studying the prerequisites of geometry, no one can claim to have learned much about water harvesting unless he or she has first understood the power of pure, prolonged, nonjudgmental, and multisensory observation. I'm talking about being open and receptive to information from each of your senses, concentrating on them one at a time and also as a living whole, and fully absorbing the happenings of your immediate environment. Later, so that you don't lose them, use these observations as the content for writing notes, taking pictures, drawing sketches, listing inventories, obtaining data, and cataloging it.

Focus on determining basic facts. Where is it warm? Where is it windy? Where is it wet? Observe all of the places where water flows on your property. Inside your home you'll likely find sinks, showers, tubs, toilets, and a washing machine.

Outside you'll discover *canales* or gutters and downspouts, maybe a sprinkler system, sidewalk or driveway edges, bottoms of slopes, culverts, water runoff ditches, and ephemeral wetlands in low-lying areas.

During all of this time that you set aside for observation (which can of course be credited to your time becoming gradually green), watch out! It's easy to impose preconceived notions on systems. In permacultural terms, this kind of imposition often leads to a "Type One Error." Such errors are mistakes in a design that prove to be incurable and irrevocable.

Every situation faces a variety of external and internal forces that should affect the design of any system imposed on it. Only with nonjudgmental observation at the outset of the design process can we expect a successful project. Whether your observations are destined to design a landscape, to plan a business, to streamline a bureaucracy, or organize an issues-based political campaign, you must first have the nerve to observe the physical terrain ahead. Otherwise you will not fully comprehend the environment within which you will be working, and you will base your decisions only on a vague and presumptuous understanding of the preexisting conditions and systems that you can only assume exist. These assumptions, to say the least, could not be counted on as accurate, and in many cases such assumptions can end up causing significant harm, property damage, and financial loss.

Be careful not to stop the observation process, limiting your awareness to your "traditional" senses. We must, as Mollison says in a slimmer book called *Introduction to Permaculture,* "sense heat and cold, pressure, stress from efforts of walking or prickly plants, and find compatible and incompatible sites" for every component of our landscape designs.

If we observe long enough, we can recognize important patterns, such as where water flows during heavy rains, where fire has passed through the site, how certain plants tend to grow next to certain other features in the landscape, how traffic patterns affect the entire environment, and so much more. All the maps, charts, pictures, and other data provided from afar can never match the power of real-life observation.

Begin by Building Soil

The place to start the physical work of water harvesting is in your soil. If your soil can retain moisture, then it will be harvesting rain for you. If your soil allows water to rush off of your property, then it is very likely that your land can and should become more absorbent.

In soil science, the bottom soil layer is called the "C horizon." It stands for the near-lifeless layer below the "B horizon," which is the layer of reasonably healthy earth just below the lush "A horizon." The A horizon, also known as "good topsoil," is what gardeners try to create every day. It's the healthiest, most permeable layer of soil. This kind of *tierra bonita* is great for plants, primarily because, almost by definition, the porosity of this thinnest layer of soil is especially efficient at harvesting precipitation.

If winds blow or rains fall and your property is literally tossed downhill, then your land, by its very existence, is contributing to an array of environmental problems, including the pollution of your local watershed. If, on the other hand, populations of insects, microorganisms, and mycelium (which can be understood, at least metaphorically, as the "roots" of mushrooms) in your soil increase with time, then your property, by virtue of its increased ability to absorb moisture, contributes to the solution of both your backyard's and your community's water problems.

In the short term, revegetation will occur, and this will typically increase your property's biodiversity. Later, as moisture percolates into the ground, as shade trees emerge and windbreaks begin to have their positive effects, and as your chosen plant material starts to thrive, your land becomes more inviting, more beautiful, more comfortable, more valuable in your local real-estate market, and more useful whenever economic times are tough. Most importantly, although some of the moisture that you harvest within your soil remains in the earth, and while a percentage of it evaporates or transpires through plants into the atmosphere, a significant portion of the water that you harvest with your soil will often find its way into local aquifers or surface water supplies.

Although the purpose of this book is to provide alternatives to the conventional water sources upon which we depend, one of the long-term effects of passive water harvesting is the replenishment of our aquifers, rivers, and reservoirs. Ultimately, this refilling of our supplies is a very important part of our water sustainability because during dry times in the future, we will inevitably need to draw water from these conventional sources, and we will want to ensure that that water exists in them indefinitely.

Mulch

One of the easiest ways to harvest rain is to spread mulch. Mulch is any material placed on the ground that builds soil. Straw, bark, pecan shells, wood chips, compost, and gravel are some of the most frequently used mulching materials.

During strong rain storms, a layer of mulch in the garden or on your landscape controls soil erosion by lessening the impact of raindrops and by reducing the powerful forces of runoff. By thwarting soil erosion in this way, a layer of mulch in the garden or on your landscape also prevents evaporation while creating comfortable conditions for the mycelia and beneficial insects that generate a healthy A horizon. Like soil building, mulching makes soil more porous and more easily able to absorb moisture in the short term, while it can simultaneously effectively replenish your watershed's conventional supplies in the long term.

Since most mulch materials are light, inexpensive, and easy to install, the energy output associated with mulching is minimal, while the payback from a water harvesting perspective is enormous. In addition, by simply spreading a 3-to 4-inch layer of mulch, you can often reduce or eliminate garden chores such as watering, tilling, fertilizing, and weeding.

You can mulch at any time of year. Thick blankets of bark laid down in fall or early winter keep root systems warm during the coldest times of year. In New Mexico or most any region, early spring is a good time to lay down a unified matrix of straw in the form of slightly loosened "books," or "flakes," of a straw bale. This acts to prevent the damaging effects of wind.

Immediately before your bioregion's rainy season is often the most important time to mulch; without mulch precious rainwater runoff is wasted and much of the moisture that might be retained in your soil is soon lost to evaporation. During any precipitation, valuable flora will wither without mulch, soil will slip away, and an easy opportunity to gradually green your life will be missed.

Another good time to mulch is during any political mudslinging season. When the trash talk in your local newspaper has become all you can bear, why not put your local rag to use in the garden? Most newspapers now use soy-based inks, so thick layers can be used as a weed barrier and thin layers can be applied as a passive water harvesting technique. Both techniques require that you put a second layer of mulching materials, such as gravel or bark, on top of the newspaper so that the newspaper won't blow away and make your front yard look like a recycling center.

One very effective mulching method is called "sheet mulching." It consists of a layer of corrugated cardboard covered by 2 to 4 inches of manure and topped with a 4-inch layer of carefully placed straw. The glue in the cardboard attracts worms, the manure provides nitrogen, and the straw supplies carbon. Sheet mulching creates a trifecta of healthy, water-absorbent soil. In addition, the triple layer of material is an excellent weed barrier, made out of the basic elements of compost: carbon, nitrogen, and beneficial insects.

Most importantly, buried cardboard creates a cool, airy, and protected medium that is conducive to the rapid spread of mycelium across a landscape. These highly productive matrices, which act like the root systems for mushrooms, are among nature's most efficient conduits for moisture and nutrient delivery. Mycelium are in some way the circulatory system of healthy soil.

Unlike a layer of compost spread on the surface of your soil, sheet mulch can easily withstand the harsh effect of the elements such as hot sun, relentless wind, deep cold, and heavy rains. Be advised, however, that fresh manure, which is a valuable fertilizer, can be too hot (too high in nitrogen) for young plants. This means the best time to sheet mulch is several months before planting time. After the sheet mulch has had weeks or months to decompose and you are ready to plant in your garden or landscape, simply push the straw to the side and dig into the nutrient-rich layer of soil that you have created with the mulching materials that have properly digested under the straw.

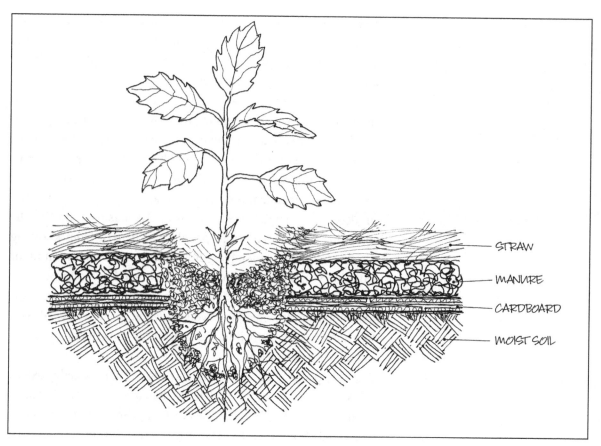

STRAW

MANURE

CARDBOARD

MOIST SOIL

Sheet Mulch

Spreading chipped plant material, bark, straw, gravel, compost, or other appropriate materials on the ground in planted areas can be essential for the success of land-restoration projects. Protecting the ground from the wind and sun helps retain moisture, so plants need less precipitation and fewer supplemental nutrients in order to survive. Keep in mind that in extremely fire-prone areas, particularly near structures, the combustion potential of vegetation-based mulches can increase your property's risk of fire damage, so it is wise to use an inert material, such as gravel, in such situations.

Gravel has another significant benefit worth mentioning: you do not need to replenish it as often as other mulch materials. Typically, this means that maintenance costs related to gravel mulch are low. However, due to the amount of energy needed for mining and transporting gravel, its environmental impact is probably no better than the extra quantities of water necessary to grow straw mulch, which typically requires significant supplemental irrigation and petroleum-based fuels for the sowing, growing, packaging, delivery, and installation of the bales to be used in your project.

Since there is so much virtual water imbedded in mulch, that is, water necessary to grow it on a farm or dig it out of the ground (not to mention the energy needed to deliver it to your garden), one could argue that mulching is a waste of water. And this would be a valid argument if our destiny were to always be dependent on food from far away. But it is, in fact, our destiny to do the exact opposite. Increasingly during the impending water revolution, people will benefit from knowing how to grow local vegetables, how to create air conditioning or insulation with a swath of plant material, and how to grow vines for privacy in our increasingly dense neighborhoods. Perhaps most importantly, healthy, well-mulched plant material will often inspire others to join the grassroots movement toward a gradually greener Earth.

Especially on degraded land above channelized erosion, a less common type of mulching manages a form of runoff called "sheet flow." According to Craig Sponholtz, president of Drylands Solutions, Inc., sometimes the best mulches consist of one layer of fist-sized rocks, deliberately placed to maximize surface contact with the soil. "Start as high as you possibly can on your piece of property," Sponholtz says, "and think of your work as being as much about sediment control as it is about passive water harvesting because it's always cheaper to control small amounts of sediment high up in the watershed than it is to deal with huge quantities of sediment downstream."

Sponholtz also uses branches and other slash-debris materials left over from forest-fire-prevention projects. Like other mulches, a matrix of sticks and pine needles will protect the soil from wind and the harsh sun. Such a covering can also help keep pedestrian traffic on pathways and away from land that you want to regenerate. An overly dense layer of branches, however, can become a nesting ground for a variety of critters, so be careful to use slash lightly near buildings. Also known as the lop-and-scatter method of erosion control, mulches made of dry branches will increase the risk of fire in any area, so be sure to separate swaths of cuttings by large, wide-open areas whenever fire safety is an issue.

For most people, mulching is the easiest way to harvest the rain. If you get nothing else from this book other than enough inspiration to use more mulch and less water in your garden, this will represent an important, gradual start toward a sustainable future for humanity.

Compost

Another simple and effective way to harvest precipitation passively is to generate your own compost—a thick, sometimes slimy blend of decomposed organic matter—and add it to your existing soil. Your resulting water-absorbent property will decrease the water needs of your plant material, reduce the sedimentation of local waterways, and help to increase local water supplies.

In Gramma's home, "forgetting the compost" was not an option. If my family had an official bad-behavior scale, neglecting to scrape your plate into the milk carton next to the sink after a meal hovered somewhere between swearing at the dinner table and using improper grammar in a crowded theater.

"Just bury me in my compost pile, so I can come up flowers," she would say with proud humility over a late-night glass of sherry. For two or three very good reasons, laws prohibit compost-pile burials, but to honor the spirit of her request, my Aunt Marjorie reverently sprinkled a small bag of her mom's celebrated black gold on top of Gramma's shiny black coffin as it descended below the dark earth.

There are about as many ways to compost as there are people who have ever tried to do so. Compost methods are the matter of legends, a secret sauce to some, the mark of a true gardener to others. Successful composting, however, usually falls into one of two categories: cold composting and hot composting.

Cold composting is best described in John Jeavons's essential gardening book, *How to Grow More Vegetables*. The beauty of his system is that you only have to turn cold piles once because of the use of thin layers of green (nitrogen-rich) and brown (carbon-rich) materials that you lay on top of each other at the outset. This makes turning the compost much less necessary. Cold composting is best started whenever large quantities of materials are available. Cold piles take about an hour each to build, but need much less maintenance than hot piles. For many people, the best part about a cold compost pile is that it digests on its own so you don't have to add to it regularly the way you do with the hot variety.

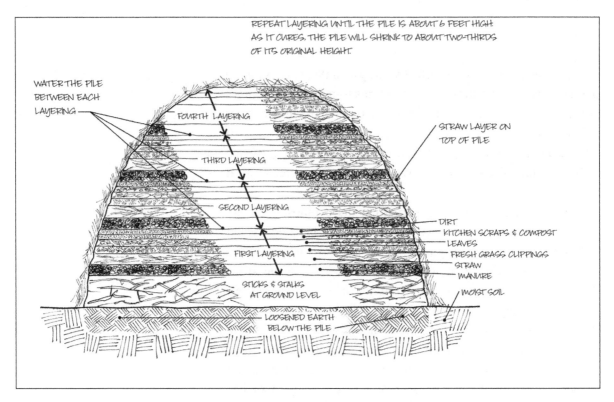

Jeavons's Cold Compost Pile

Hot composting is the more common form of composting and works best when materials are added gradually over time. Jeavons notes that hot compost can also get too hot and can even burn certain microorganisms. Traditionalists like my grandmother tend to argue that cold compost piles will not burn up weed seeds as effectively as hot compost piles, and who wants weeds in the garden?

In my experience both methods work well, so the vital question isn't whether you gather up all of your materials at once and create a Jeavons-style cold compost pile, or whether you gradually add to a pile or compost bin that ferments organic material over time. What's important is that you pump up your rain harvesting by using compost on your property.

Gramma chose a perfect site for her three-bin system: in deep shade, protected from wind, near the kitchen, close to a hose bib, and under the roofline of her garage, where precipitation would run off into the piles during every storm event. The first bin received incoming kitchen scraps and yard wastes. The middle bin held half-finished compost. The last bin provided compost ready for use.

Her bins were made of wood slats and wire fencing. This allowed air to get into the pile, while it discouraged rodents. She watered and turned her piles regularly, and her product was black humus, teeming with worms, thriving with life and pounds of potential.

Turning your compost pile is great exercise, and, as such, this chore will be revered in gradual-greening circles through the ages, but my friend Tom Knoblauch designed an efficient compost-turning tool that saves time and reduces the physical exertion associated with this laborious, yet glorious, task. "For years I've built my piles around one of those perforated, 4-inch-diameter plastic sewer and drain pipes," he told me over strong black coffee and a quick, back-of-a-napkin sketch. "First, you cap the pipe at the bottom, then you shove eight or 10 short pieces of rebar through the perforations, and make sure the diameter of the rebar isn't a whole lot smaller than the diameter of the holes in your pipe." Whenever he wants, Knoblauch can turn the top piece of rebar and all of the other pieces turn with it. "This churns the compost perfectly," he went on. "Plus, my tool lets me water my compost more thoroughly because I can pour water into the top of the pipe, and it'll seep out the perforations in the pipe, getting the entire pile nice and moist, top to bottom, with minimal water."

Every gardener's favorite ingredient in compost is animal manure. Every Mother's Day Grampa always knew what to get Gramma. It wasn't jewelry or a box of chocolates, it wasn't a family dinner in the big city, and it certainly wasn't roses since Gramma's garden would soon be full of them. Grampa knew to call his friend with the pickup truck who lived near some horse stables. More than anything for Mother's Day, Gramma wanted an overflowing truckload of animal waste to make her compost pile as productive as possible.

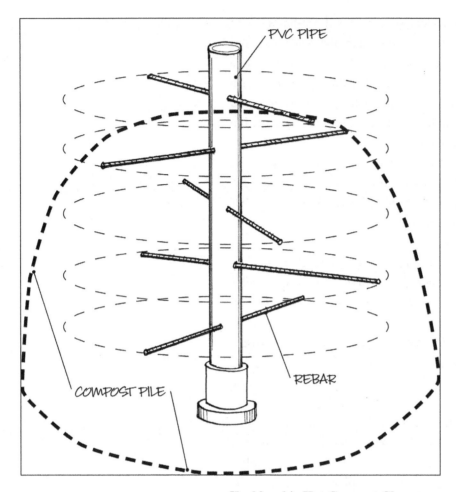

Knoblauch's Hot Compost Churner

My excrement of choice is chicken. Chickens are easy to keep alive, and they provide one of my favorite foods, eggs. Chickens also can keep weeds down in and around the garden as they simultaneously keep some insect populations at bay. But one of the best reasons to have chickens is that, with their urine and excrement, they provide awesome sources of fresh, organic nitrogen.

No material will better improve dry, nutrient-deficient soils than homegrown compost. For best results, use a trowel to hand dig or scratch your compost into the soil. Using compost as a bottom layer of mulch, covered with any of the other mulches described in the previous mulching section, is also an effective technique. I have also had success by placing kitchen scraps in small quantities under the mulch in my garden. Just lift up a thick chunk of straw or bark, dump small amounts of leftovers on the ground and then place the mulch back on top. This eliminates the need to build an actual pile and cuts back on the turning and watering that go with maintaining a hot compost pile.

It is important to avoid adding to your compost pile (or placing under your mulch) significant quantities of meat and dairy products, oils, fats, poisonous plants, or plants infected with disease. Pernicious weeds and significant quantities of cat and dog manure are also no-no's.

Consider covering your compost pile with a thick straw mulch, add some red worms (which can be ordered over the Internet or, better still, bought locally), and do not forget to keep the compost as moist as a wrung-out sponge. Finally, don't feel guilty about watering it. If used properly, compost will ultimately help you save water in your garden.

Meanwhile, keep in mind that it's also relatively easy to divert liquids to your compost pile before they ever reach the drain of the kitchen sink. Unfinished beverages, vegetable-washing water, some dishwater (avoid raw-meat juice), and leftover liquids from cooking can all be poured into your kitchen's food-scrap container for subsequent dumping on the compost pile, and this reduces your need to add water to the pile.

For those who want or need to buy and/or import compost, look for local sources first. According to fellow permaculture teacher Jim Brooks of Soilutions in Albuquerque, New Mexico, "When purchasing compost, look at the ingredients. Cow and steer manure tend to have high salt content. Peat moss comes from nonrenewable bogs that play a big role in the stability of our atmosphere. Plus," he smiles while standing in front of his booth at an Albuquerque watershed conference, "peat has no available nutrients to speak of." Another unsettling ingredient often found in compost is sewage or what we now often call "biosolids." "Biosolids are used as a fertilizer all over the world, but it's generally high in metals and you just have no way of knowing what's in it," warns Brooks, "so Soilutions avoids it categorically." His company's "Premium Compost" is approved for use by organic gardeners and farmers in New Mexico. Made from "a wide variety of recycled plant materials such as leaves, grass clippings, trees, brush, cactus, weeds, horse manure, and stall bedding," his mixture is low in salts and decomposes at temperatures in excess of 135 degrees Fahrenheit.

Who's the Jim Brooks in your neck of the watershed? If you have poor soils and you are starting a large project, you will want to find out. These days this can be as easy as reading ingredients on a Web page or sending a quick e-mail or making a brief phone call.

Grow Food

Growing edible plants makes sense on many levels. With produce prices skyrocketing, it can often be an inexpensive way to provide fresh, healthy organic food for your family. As we come to this point in society's evolution not only of "peak oil" but also of "peak junk food," a tomato plant in a window, a patch of lettuce under some dappled shade at the backdoor, herbs for health and flavor, or a fully fledged food forest surrounding your home will provide pleasurable hands-on learning experiences for people of all ages.

The biological, chemical, and physical relationships among water, roots, soil, leaves, and air are all easily accessible to any open and careful mind. How much water? That's a biological question. How much sunlight? Photochemistry. How much heat? Pure thermodynamics.

Gardens and a comfortable landscape in general are, as I've said, great for your quality of life and long-term property investment. For a brief moment, we should also be candid and look at the worst-case scenario. If times ever got extremely tough, the patch of land with the improved soil will be worth much more than a neighboring piece of property that has poor and/or compacted soils.

If for no other reason than to understand the sheer amount of water that a modern diet needs, growing your own food is a responsible, empowering activity on a human level. Just like the water-drinking exercise found in this book's introduction, eating what you have grown can have a similar inspirational effect of giving a boost to consciousness and spirituality. There is something physically and metaphysically invigorating about eating food that you have been responsible for growing.

Perhaps this is why fruits from your own garden tend

to taste better than anything you could ever buy on a plate or in a package. Your body clearly enjoys it, so why shouldn't your soul? Take, eat, and, yes, use as much water in your garden as you need in order to generate your own backyard abundance.

Every calorie of food you create on your own property reduces the huge number of calories necessary to produce every other kind of food unless your neighbor happens to be a farmer/salesman of a wide variety of crops. The typical American carbon footprint is made up of four touch points—transportation, housing, industry, and agriculture—so one important direct-action tactic to combat global warming is to decrease your dependence on the mega-food corporations that need to generate tremendous energy in order to keep you alive. Our domestic and imported food is often the result of high-energy inputs such as fertilizers, pesticides/herbicides/fungicides/miticides, gas- and diesel-powered machinery, wasteful irrigation practices, exploited labor, inefficient transportation systems, titanic energy grids, worldwide marketing maneuvers, dangerous technology, and a mostly untrustworthy financial colossus, so the environmental impact of growing a garden is worth the water consumed in all but the most extreme cases in which drinking-water supplies are critically low.

Our combined human impact is more than an Achilles heel; it is a yoke we must reduce if we are to create a living, breathing planet. In a 2007 *Rolling Stone* interview, James Lovelock, author of the Gaia hypothesis, put it this way:

> There are nearly 7 billion people on the planet now, not to mention livestock and pets. If you just take the CO_2 of everything breathing, it's twenty-five percent of the total—four times as much CO_2 as all the airlines in the world. So if you want to improve your carbon footprint, just hold your breath. It's terrifying. We have just exceeded all reasonable bounds in numbers. And from a purely biological view, any species that does that has a crash.

If Lovelock is correct, we have created an overpopulation problem to worry about, which is forcing Gaia, the planet's spirit, to respond with severe climatic shifts. This will

My hope is that if we learn to harvest water locally in this generation, we will build a strong foundation for growing food locally in the next generation.

increasingly threaten life as we know it. Our work, for those who recognize the looming threats, must be about rebuilding systems in a manner that is aware and, in practice, reduces adverse environmental impacts. Food production that moves away from mega-agribusiness is a vital part of this solution. This will mean—sometimes more than rainwater harvesting—that food production at a local level will be essential to our survival. My hope is that if we learn to harvest water locally in this generation, we will build a strong foundation for growing food locally in the next generation.

We might consider with new insight why backyard gardening, farmers markets, organic growing, heirloom-seed saving, and decentralized water-saving systems all deliver more potential for our future than today's "norm"—a petrochemical-dependent, water- and soil-depleting system, corporate-farm regimen based on fragile currencies. It's time for a change, and the well-watered vegetable garden or food forest may be the best place to strengthen any meaningful revolution.

Cut Swales

Swales go back to ancient times but more recently have been rethought and introduced with permaculture. By contouring the land to increase water infiltration from rainfall and snowmelt, you can capture significant moisture using a simple technique called an "on-contour swale."

This type of ditch is dug along points on a line that are all at the same elevation above sea level. The excavated dirt is carefully placed and tamped on the downhill side of the ditch in the form of a berm. In order to prevent runoff from pouring out the sides of a swale, each end of the berm must curve uphill, making a vessel, or trough, in the ground.

During and after large precipitation events, on-contour swales become intentional miniponds. Instead of eroding the soil further, the harvested runoff either percolates deep underground to become part of an aquifer (or sometimes a subterranean waterway), or it remains in the soil for use by plants in the newly created, wetter microclimate.

Swales can also be dug "off-contour" in order to divert water toward a tree, shrub, or garden bed, but if they are too big or steep, off-contour swales can cause more harm than good.

This damage can come in the form of soil erosion associated with significant storm events. Large quantities of runoff can wreak havoc, so in most arid-land cases in particular, it's best to dig simple systems of several on-contour swales rather than one off-contour swale.

In the case of on-contour swales, points on top of the berm should be level unless you decide to install a spillway for excess runoff. With a level berm, runoff can overflow evenly and gently over the entire berm (hopefully into another swale). If your berm is not level, during a major storm event the minipond of water behind the berm will find a low point on the seemingly harmless mound of earth and subsequently blow a wide hole in your berm, leaving a muddy mess in its wake and a lot of wasted water.

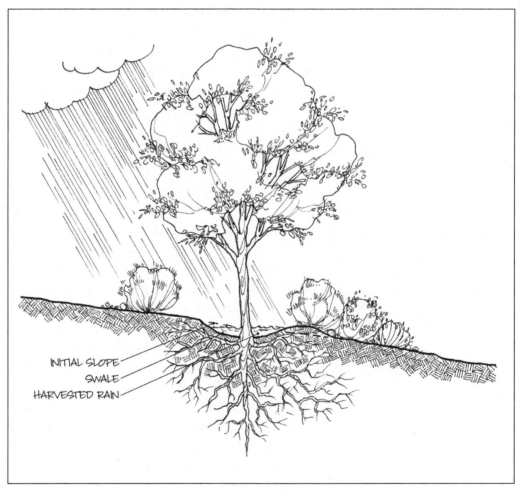

An On-Contour Swale

If you prefer not to worry about leveling your berms (and making sure they remain level over time), make sure you determine how excess water will escape your swale. Significant erosion can occur whenever you store large quantities of water on a piece of property. Appropriately placed spillways can be lined with turf or other kinds of plant material, though spillways are often lined with the rocks that are dug up during the process of creating the swale. If your soil isn't rocky, you may want to import rocks, but before doing so check to see whether an appropriate substitute material exists somewhere in the vicinity of the swale. Leftover pieces of flagstone, brick, river rock, chunks of concrete, and other hardscape materials can be a great resource for building spillways in swales.

Digging swales requires muscles and shovels—and a few accompanying instruments. Start with any of the following: a global positioning system, a transit, a water level, an A-frame level, or even a carpenter's level. Using your chosen leveling device, place flags, stones, or shovel markings in the existing dirt to demarcate a level line (known as a contour line) along the ground. Dig along the line and place the excavated soil on the downhill side of your wide, shallow trench. Next, rake, shape, and gently tamp the berm so that water can percolate into the ground during precipitation events. Finally, sow appropriate grass and wildflower seed, lay mulch, and plant drought-tolerant plants in and around the swale.

It is critical that the berm bends uphill at either end of the swale. These two curves, called "wings," ensure that all of the available runoff is captured. Think of these wings as the bow and stern of a canoe. Without the two curves up and away from the middle of the keel, such a vessel would look and function more like a raft—able to hold much less loose cargo. Your swales will harvest much less water if you neglect to finish them with curved berms directed uphill.

Swales are often best lined out in a fish-scale pattern such that as the highest swales on your slope fill up with rainwater, they spill naturally into the swales below them. This allows for the kind of water infiltration that could help local aquifers refill and local soils come to life. If your plant material is spread out widely on a slope, you can modify the fish-scale pattern in such a way that the majority of your runoff water is first captured in off-contour swales that deliver a large surface area's worth of precipitation into a planting area. This system is known to permaculturalists as a net-and-pan water harvesting system.

Another helpful technique is to install permeable berms made of organic material along the contours of your slope. Jute wattles, rows of rocks, straw bales, straw books (the distinct square sections that make up a straw bale, also known as straw-bale flakes), or almost any other permeable materials can be installed along the contour of a slope in the much the same way as an earthen, on-contour swale. These permeable berms also prevent erosion and harvest precipitation.

Permeable berms are used instead of earthen swales in two types of situations. When a slope is too steep for earthen swales, permeable berms can be installed because they are typically nailed to the ground with wooden stakes or rebar. This prevents the berm from sloughing off the slope in the way that an earthen berm would.

Another occasion for choosing a permeable berm over an earthen swale is when your native soil is essentially starved of organic matter or completely compacted by construction. The importation of organic matter often serves as the most efficient way to ensure that moisture is stored in the soil. If the importation of such material is not cost-efficient, native rocks from your property lined up along contour can have a similarly positive effect, but be careful not to remove too many rocks such that erosion is created elsewhere on your land.

Swales can direct water precisely to the root zones of the plants associated with them and therefore should be understood as one of the most efficient passive water harvesting techniques available. Swales should be relatively narrow in order to prevent both the loss of water due to evaporation and the loss of plant material due to oversaturation. That is, if you plant in the middle of a too-wide swale, you run the risk of overwatering plants and trees during and after major storm events.

You can easily install swales properly as long as you remember the permaculture principle discussed in the first section of this part of the book: start small. Little projects allow you to learn from your mistakes and improve upon your work in between storm events. Equally important is the fact that smaller projects prevent the unnecessary squandering of energy and resources. Still, the true emphasis of this principle is on the "starting" part. If you start and never stop harvesting rain with swales, you will soon become a master of the art and science of swaling, with a nod toward what works most productively in your particular part of the world and with careful regard to the unique set of goals for each specific project you encounter.

Carve Keylines

Although large-scale farming and ranching techniques are not the province of this book, any discussion of passive water harvesting would be incomplete without an explanation of Percival A. Yeomans's "keyline" system. In *Water for Every Farm*, several other books, and countless articles, Yeomans, his sons, and their followers describe a precipitation harvesting system that uses a combination of on-contour and slightly off-contour swales and/or plow cuts that store water in the immediate soil as well as behind dams located at the "keypoint" in any primary valley, gully, or *arroyo*. This all-important place is found where the gentle, flatter portion of a primary drainage becomes decidedly less gradual, if not precipitous.

Keypoints are typically the most efficient and productive locations for storing water behind the relatively low-cost, earthen dams that Yeomans's informal school promotes. Each dam usually has a diversion channel or conduit associated with it, and these typically use gravity to flood-irrigate neighboring fields long after the storm event that filled the dam has passed. Keylines are most often thought of as the contours that lead toward the keypoint when runoff is filling up behind a dam, but they also, in a very effective manner, manage overflow from dams at keypoints during large storm events.

According to J. MacDonald-Holmes in a paper titled *The Geographical and Topographical Basis of Keyline*, keyline planning envisages first the discovery and then the development of natural renewable resources of the landscape in order to produce a "state of balance" that will be in conformity with land shape, climate, and soil. The basic idea is to make nature assist the farmer instead of his engaging in conflict with nature, to his economic loss. What the farmer and rancher/grazier need is not an unstable "out of balance" landscape, but a permanent and improving one. Keylines can bring equilibrium, MacDonald-Holmes says, by overcoming the "natural flow pattern of water on land, holding the water on ridges longer, and thus evening up the moisture content of the soil."

The system also requires a particular plow attachment that cuts slits in the ground without tearing it up. This allows for the infiltration of runoff into the topsoil, the percolation of water into the local aquifer, and a significant reduction in water loss due to evaporation—while preserving the majority of biotic life that exists in untilled soil. This is particularly important in low-tech, flood-irrigation situations.

Often touted as a drought-proof agricultural method, the system has been most successful in non-arid environments, especially wherever slopes are considered too steep for conventional farming and ranching practices. Both Mollison and David Holmgren, the lesser-known progenitor of the word *permaculture*, promote Yeomans's work. In a paper called "Development of the Permaculture Concept," Holmgren writes that "keyline provided an ideal broad scale land development framework within which more intensive permaculture systems could be applied." Although keyline systems have their critics, it is easy to see why conventional agriculture would shift toward this precipitation-based system because it can require few, if any, fuels and little, if any, supplemental irrigation—at least during wet years.

Dam Your Runoff

Erosion-control structures installed perpendicular to the flow of ephemeral watercourses are often called "weirs" or "check dams." When riprap rock is wrapped with wire gabion cages and Reno® mattresses, the structures are often called "gabions." Gabions have been around since at least medieval times and allow water to pass through them, but they retain moisture, sediment, animal manures, insects, silt, and many more essential elements of healthy soil.

Designed to be relatively permeable, these damlike constructions also hold back seeds. If built correctly, they can create an entire system of terraces in an otherwise-polluting drainage ditch. The land around these structures becomes a minioasis that steps gently down the bottom of the gully. Although it can take a while to see the positive effects of any newly created gabion, check dam, or weir, after one has been installed, increased biodiversity begins to move in as early as the first rain event.

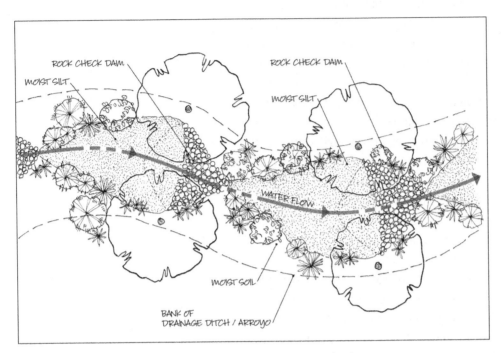

Check Dams (top view)

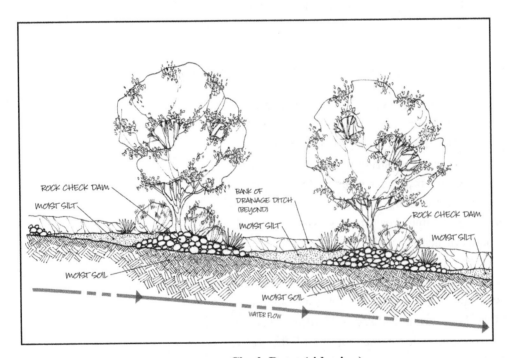

Check Dams (side view)

Even though runoff is allowed to take its natural course, it is slowed down significantly by check dams and associated sediment. The reduction in water velocity not only encourages a much greater percentage of runoff to percolate into the neighboring soil, but the end result also includes more percolation into local aquifers, less pollution of local watercourses, and a sharply increased microclimatic variety in the vicinity of the dam.

When permeable dams are made of sticks and branches, they are called fascines (in Europe) or wattles (in North America). Wattles can also be made of long fibrous bags, similar to burlap and often made of jute, coconut, or a plastic mesh, that get packed tightly with straw, jute, coconut, or other natural fibers. These human-sized, sausage-shaped materials are typically nailed into the ground using wooden stakes or rebar in such a manner as to ensure that the center of the wattle is lower than its sides. This allows for runoff to take its natural course at a slower pace through the channel.

When permeable dams are made of loosely stacked rock, these are usually called check dams. Like wattles, the middle of the dam should be significantly lower than the sides of the banks of the watercourse, but rock check dams should also run relatively long and gradual in the direction of the ditch's flow. This allows for runoff to be slowed down by the initial impact of the rocks, while it also helps to dissipate the energy of the runoff as it crosses over the check dam. Since check dams are not nailed to the earth in the way that wattles are, a long check dam (angled parallel to the flow of runoff in the ditch) also means more mass and, hence, less chance of being demolished by a major storm event.

Sow Seed, Plant Trees

Since harvesting precipitation into the soil is the goal of any passive water harvesting system, a word must be reserved for the all-important work of sowing the seeds of native pioneer grasses and wildflowers wherever one finds denuded soil. The meadows that grow out of this kind of seed sowing can prevent the sedimentation of local watersheds, help recharge local aquifers, and, ultimately, as meadows cede space to shrubs and trees, these mere seeds create forests full of biodiversity.

Like mulch, grasses and wildflowers provide shade, insulation, nutrients, and wind protection for bare ground. But they also actually hold the soil together with their root systems. All of this provides a perfect environment for the kinds of biota that make the soil more absorbent, more able to harvest the rain.

Getting grasses and wildflowers to grow without supplemental water is extremely difficult in the desert. Poor soils, strong winds, ravenous wildlife, an incessant sun, and little rain make seed germination a serious challenge. If we work within patterns found in nature, we can increase our chances of success.

Fortunately, with the help of modern meteorology and a basic knowledge of what seeds need, you can reclaim your land in any environment above sea level and below the tree line. It just takes a pinch of preparation, a dash of educated guessing, and a little luck.

The first step in being prepared is to acquire appropriate seed. A diversity of native species is best because each species has a particular combination of conditions to which it will most vigorously respond. Santa Fe–based Plants of the Southwest carries diverse grass mixes for much of the southwestern region of the United States. Other sources can be found at your local plant nurseries and on the Internet. For northern New Mexico, my favorite mixes are Dryland Blend and Sandy Soil Stabilizer, as well as an excellent wildflower mélange called High Plains Piñon-Juniper Mix.

Seed diversity is especially effective when mixes include warm- and cool-season species. This ensures that more bare ground is covered throughout the year. A wide seed palette will also create a more productive and beautiful landscape—not only by being robust and diversified, but by revealing nature's various contrasts of color, form, and texture.

Diversity of species also benefits the land because groups of companion plants tend to grow together in distinct patches, which then protect each other from harsh winds, cold nights, and hot daytime temperatures. Other than in cases where a particular aesthetic or need is to be considered, both wildflowers and grasses should be sown together in the area that you wish to revegetate.

When attempting to cover barren soil quickly with

vegetation, the main reason for diversifying your seed palette is that such diversity increases your chance of success. Especially if you plan to depend on Mother Nature for your moisture, a wide variety of species translates into at least a species or two surviving under almost any naturally occurring set of physical conditions.

To improve germination rates, don't cover your seeds with too much or too little soil. Seeds generally like to be buried as deep as their width—that is, not very deep. Simply raking an area after seeds are sown works pretty well, especially if mulch is specified to be added afterword.

Seeds also can benefit from mulch—but not too much. In order to retain moisture and protect your seeds from wind, add a thin layer of straw, bark, or compost. A natural white powder called "binder," or tackifier, can usually be purchased from your local nursery and should be mixed with your seed before sowing.

If you truly do not plan on watering the area that you sow, you will need daily access to a reliable five- to ten-day weather forecast because your aforementioned, all-important luck will depend in part on the accuracy of your ability to forecast the weather. Be prepared for one of those weeks with a large number of rainy days. When you see that one of those wonderful weeks has finally arrived, don't delay: start sowing as soon as you see those first clouds forming.

Getting pioneer species of grasses and wildflowers to germinate is an important part of building your soil, but don't forget to include native and appropriate trees, shrubs, perennial flowers, and edible plants in your landscape and garden guilds. These all take water to establish, but if properly chosen, they usually come with benefits that far outweigh the need for the water they consume.

Such a food forest can be seen as an appropriate ultimate goal for any wildflower meadow or grasslands. Open areas, by their nature, represent vacuums that nature is known to abhor. At some point, shrubs and trees will begin to encroach on any meadow or grassland. When pioneer species settle an area, they cultivate the earth for future generations of life that come in with entirely different habits of growth. This is what permaculturalists call "succession," and whenever you can work with this natural process, rather than against it, you will probably be more efficient and productive. This is especially true in the worlds of revegetation and reforestation.

Water at the Root Zones

For years my landscape-installation company planted a whole lot of polyvinyl chloride (PVC) pipe in the ground. We weren't proud, but we justified these installations because our deep-pipe irrigation technique saved so much water. At a time when much of our competition didn't know how to comply with drought restrictions and keep new plantings alive, the technique I had developed from a decade-old David Bainbridge article in the *Permaculture Drylands Journal* helped make drought years as profitable as wet years for Santa Fe Permaculture's installation crews.

We tried to use as much scrap pipe as possible. We knew PVC releases high levels of dioxin, a nasty carcinogen, during the manufacturing process, so we avoided buying it. Plus, any old scrap pipe would do. It just had to be at least a foot long and 1.5, 2, 3, or even 4 inches in diameter.

In the old days, first we would cut a 2-inch diameter pipe into as many 18 to 24-inch lengths as we needed. We'd then cut slits along one side of the pipe at 2 or 3-inch intervals, all the way down. Next, we would dig a slightly wider hole for the particular species of plant that we planned to plant. Then we would stick the perforated pipe next to the plant. Finally, we'd backfill around the plant and the pipe and fill the pipe with gravel. As long as someone secured a drip irrigation emitter into the pipe's highest perforation near the surface of the soil, the appropriately chosen newly installed plant material would almost always survive during intense droughts even at times when the City of Santa Fe imposed a once-per-week watering regime. (Of course our summer monsoon rains sometimes helped, but new plants in our area typically need to be watered once per *day* for the first couple of weeks after planting and then every other day after that for most of their first growing season.)

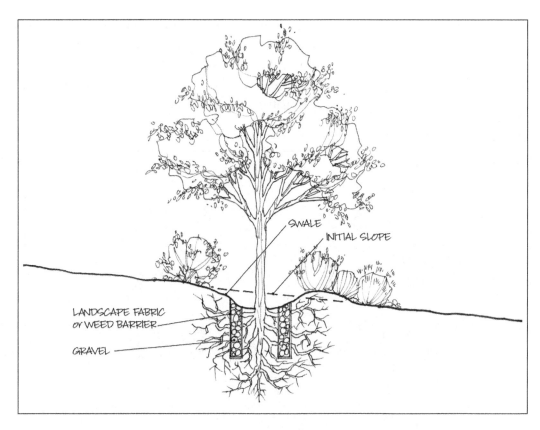

Baker's Rock Tubes

One day I met one of the world's great optimists, fellow Santa Fe permaculturalist Reese Baker, president of The Raincatcher, a Santa Fe landscaping firm. He explained the simple concept of temporarily reusing one pipe over and over. "After you pour the gravel in your mostly buried tube, dude, you just carefully pull the pipe out, and you're done!" Baker's eyes were wide and blazing. "Plus," the lifelong classical pianist slowed down almost to an adagio, "No *P, V, C*!"

Phew! I said to myself. *How nice of Reese not to throw a "Duh?" in there at the end.* What he said was so clear and so simple that I realized at that moment that I would never prescribe deep-pipe irrigation again. I'd specify Baker's rock tubes evermore.

In the same way "books" of straw, that is, those similarly sized square slices that make up a straw bale, if buried vertically next to a plant can also significantly reduce the water needs of plant material. Just be careful not to expose root systems to too much air. One pipe, tube, or book per plant is sufficient in reclamation work, but within formal courtyards two or three evenly spaced pipes or books are worth the effort because they ensure that the plant material will grow with reasonable symmetry.

A Santa Fe Drain

We've also used cardboard boxes filled with pumice or even holes filled with site-collected fist-sized rocks. These pockets, which are also what I describe as Santa Fe drains (below), are wider than rock tubes, so these can be shared among a number of plants by installing a pocket in the center of a guild of plants. These minireservoirs work especially well in the depressions of on-contour swales. In keeping with the permaculture principle that every element in a system should perform more than one function, author/teacher Bainbridge suggests that about 6 inches of pipe stick up out of the ground in land-restoration projects and this will serve as a built-in stake for securing wildlife-prevention fencing around individual plants or even larger guilds.

The concept behind deep-pipe irrigation dates back to a centuries-old strategy known as the "buried *olla*" technique. Plant material can be planted around unglazed terra-cotta pots as if the pot were one large rock tube or cardboard-box pumice pocket. Runoff during and after storm events slowly percolates toward the root zones of your plant material out of all of the natural pores in the pot or box.

Sometimes local garden and/or kitsch shops are happy to donate blemished, unglazed terra-cotta pots that they can't sell. Years ago when we designed a community garden for the local Food Depot, a Santa Fe food bank, we were fortunate to get Jack-a-lope, a local garden center, to donate a number of large pots that weren't going to sell because they'd been chipped or scratched. Since none of these pots came with lids, a drowned lizard and two dead mice taught us that these minicisterns need to be covered securely with some kind of cap, such as a scrap of thick flagstone or a heavily weighted-down piece of wood.

All of these techniques—deep pipes, rock tubes, straw books, pumice pockets, and buried *ollas*—make supplemental watering much less necessary because significant quantities of rainwater are funneled directly to the point where they are most needed: at the root zones of your plant material.

Harvest the Air

In 2002, the last year of a 23-year study, New Mexico and North Dakota tied for second place, behind Alaska, in "most per capita hypothermia-related deaths," according to *The Mortality and Morbidity Weekly Report*, published by the Center for Disease Control. Although the story was never a hot topic in any form of media, it transitions smoothly into this section's topic: harvesting condensation from ambient air.

By definition, Alaska and North Dakota are ridiculously cold places, so seeing them in the top three hypothermia-deaths list made sense. But, with New Mexico's close relationships to the Chihuahuan and Sonoran Deserts, my state has a reputation of being not merely hot but often brutally so. Surely, even if one disregards the quantities of hot and hotter chile grown all over New Mexico, my neighbors in the Land of Enchantment will agree: we live in the home of an unforgiving sun.

By the time October rolls around most locals know that deadly cold weather can set in, and what many tourists and newcomers do not know is that daily temperatures swing tremendously in any high-altitude desert. Our thin air, sharp changes in elevation, sparse vegetation, and lack of proximity to a large body of water make rapid, 40-degree temperature shifts quite common. The bad news is: these conditions are perfect for hypothermia. The good news is: this phenomenon is perfect for harvesting water in the form of condensation.

Native Americans up and down the Rio Grande Valley piled rocks around saplings planted in holes or on swales along hillsides in order to take advantage of this extreme temperature-change phenomenon. The rocks would heat up with their daily dose of obdurate sunshine, and by dusk droplets of moisture would begin to condense and even precipitate under each stone.

My friend, client, and fellow permaculturalist Mary Zemach uses a waterless-tomato technique that exemplifies a modern application of this concept. Zemach clips heavy black plastic sheets to the insides of strong, rigid wire cylinders that stand 3 feet tall and 2 to 3 feet in diameter. In autumn she fills these columns with leaves. In the spring she comes back with a tomato plant (started in her basement) and adds a few shovelfuls of soil around its roots as she buries most, if not all of the plant, with leaves. Like the rock piles of the ancient inhabitants of the Rio Grande Valley, the plastic (and associated leaf mass) heats up daily. Then it cools each night, leaving behind, in the leaves, plenty of moisture for use by Zemach's tomatoes throughout the day.

"Nature takes over from there," she says, over a salad she made us from her garden. "I never water them directly, but about six times a year, they get some overspray when I'm watering other things in the garden."

Another way to harvest water from air is to plug in an appliance called an atmospheric water generator, a machine that creates moisture out of ambient air. This new technology is designed to create enough drinking water for your household. According to one product's website, www.myxziex.com, "the air intake is filtered, the water is internally circulated, UV treated and passed through carbon-block filters, remineralized, and then chilled for maximum cold taste and freshness." Able to produce more water as the humidity in a room increases, these appliances make a great deal of sense in wet regions where water quality is an issue or in drylands as a supplementary source of drinking water that leaves more rainwater for the garden.

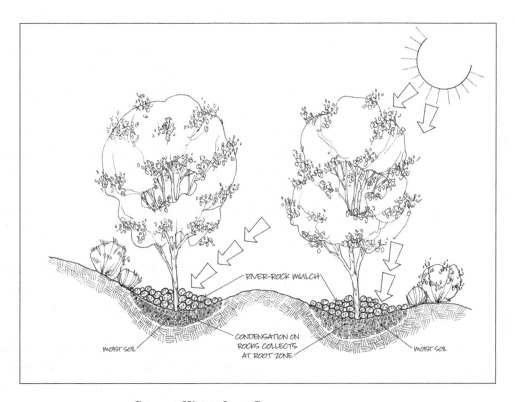

Squeeze Water from Stones

It is true that these plug-in water makers, technically, are not passive water harvesting techniques. They involve moving parts and require electricity, but they do belong in a section about harvesting water from air not only because they use the same phenomenon as waterless tomatoes, but also because there is something so mellow, so passive, about the concept of plugging a condensation-creation gadget into a standard outlet next to your fridge and having a freshwater fountain at your disposal. No plumbing connections, no water bills, no plastic bottles ever again.

Although this concept is based on age-old, dryland-farming techniques, the ambient water-generator technology is newly emergent, so I am not yet ready to vouch for it. Instead of the five-gallon plastic bottles that we reuse for our drinking water, in an arid environment I would worry about the increased number of plastic lotion bottles that our family would use to keep our skin moist if one of our appliances demanded moisture from our decidedly dry, mountain air. Already in the winter, that lotion bill goes up at about the same pace as our use of the woodstove.

Cloud seeding is an ever-popular way of attempting to increase (or decrease) the amount of precipitation in an area. Since it requires an airplane and a relatively high level of biochemistry, we won't focus on this tool of passive water harvesting, but it is worth mentioning the use of frozen carbon dioxide (dry ice), silver iodide, salt, and other particulates in an effort to condense and precipitate moisture. There is skepticism as to how well the technology works, and environmentalists warn that too many hygroscopic chemicals released into the air could create negative consequences for the land and people below. In spite of the fact that there are as yet untackled legal questions whether such seeding should be seen as an act of theft from downwind neighbors, at least nine western states have active cloud-seeding operations with scores of millions of dollars spent every year.

I'm very wary of cloud-seeding technology because it's a very expensive and inexact science with potentially harmful known and unknown side effects. In addition, I believe that in this increasingly thirsty world, whoever owns the skies with its air force should not necessarily own the weather, the clouds, and storms over the people down below. The natural right to own one's own native waters is already being fought over and determined in the main by megaindustrial powers, so we should avoid supporting the hubris of a military-industrial complex that would thoroughly enjoy taking over the skies in an attempt to control the activities of people on the ground.

Dig Drains

Over the years, I have noticed confusion surrounding the constituent parts of a French drain, so here I will make a simple distinction. Many people, myself included, think of a French drain as simply a hole filled with gravel, rocks, or rubble. Such drains control erosion, replenish local aquifers, and harvest runoff for nearby plants. (They look pretty much like the illustration of a pumice pocket, above, but without the cardboard box.)

Unfortunately, other folks think that a French drain is much more than this. They believe that a piece of perforated drain pipe, often with a stockinglike piece of landscaping fabric wrapped around the pipe, must be part of a French drain. Typically, this pipe runs laterally through a trench filled with gravel, rocks, or rubble. The pipe carries runoff to an intended place at the end of the pipe, while it may or may not allow for some localized percolation along the trench. (This type of drain looks much like the picture of a pumice wick, below.)

Although it's sometimes hard, I do my best to be a nonconfrontational guy, so I'll give in on the whole French-drain concept and invent a new term for the simple rock-filled excavations. From now on I'll try to use the term *Santa Fe drain* whenever we plan on leaving the perforated pipe out of our trenches or holes that we plan to backfill with aggregate. Although it's a new term, it's an age-old water harvesting technique predating not only Santa Fe and France but all forms of written law. Since, at least in the vernacular of water harvesters whom I know, no term has stuck for a hole filled with rocks, my thought is this: Why not give "Santa Fe drain" a try?

You might think that a hole filled with gravel in a place like Santa Fe wouldn't provide much water for the root systems of plant material because the water would drain through the rocks and into the sandy desert below. Because of some much-prevalent sand-dune imagery, many people erroneously think of deserts as sandy places. It is true that drylands have plenty of sand, but many also have significant clay content in the soil. The dysfunction in such places exists because there is no organic material to integrate the sand with the clay, and the main reason that organic material is absent is that there isn't much water in the soil.

Resist Gravity

It's sad to think of where all of that awesome precipitation goes after a large storm event in an arid land. Most of it careens off roofs, roads, and barren earth, and moves much too fast for any plant material to take it in. Ironically, plants often get pummeled and are sometimes even destroyed by the very thing they need most: water.

Why not store this resource in a no-maintenance *pumice wick*? These invisible "sponges underground" are harder to install than Santa Fe drains—but not by much. Over fifteen years ago, Derk Loeks, a passive water harvesting and erosion-control shaman in his own right, taught me the technique. Although some of the details have changed, the basic idea is still the same.

To begin, fill a trench with a porous, hygroscopic stone like pumice (or crushed recycled glass) and divert roof or road runoff into the trench. After laying a barrier, such as a thick layer of newspaper or a strip of weed barrier (also called "landscape fabric"), on top of the pumice, cover the trench up with earth. The goal is to harvest runoff from any impervious surface. When plants are planted on either side of the wick, they are given a supply of moisture at their root zones for weeks or even months after every good rain.

In the paragraphs below, you will find some basic guidelines for building pumice wicks in association with a roof, but this process can be easily adapted to harvest water from driveways, roads, and patios, too. Note that it is important to use a light, porous stone so as to maintain the spongelike effect of the wick, as opposed to the drainlike effect of gravel. One ecological substitute for pumice would likely be a recycled-glass product being developed by Santa Fe–based Earthstone International.

First, under a downspout dig a hole big enough for a small box drain and direct the box drain's outflow pipe in the direction of your wick. Under flat roof *canales*, you will need to collect precipitation in a funnel drain. These resemble the box-drain systems for downspouts, except that they are installed below ground level and are covered with gravel. A 6 foot by 6 foot piece of heavy-duty plastic, such as pond liner or shower liner, is used under the gravel and above the box drain in order to funnel water toward the box and into the wick.

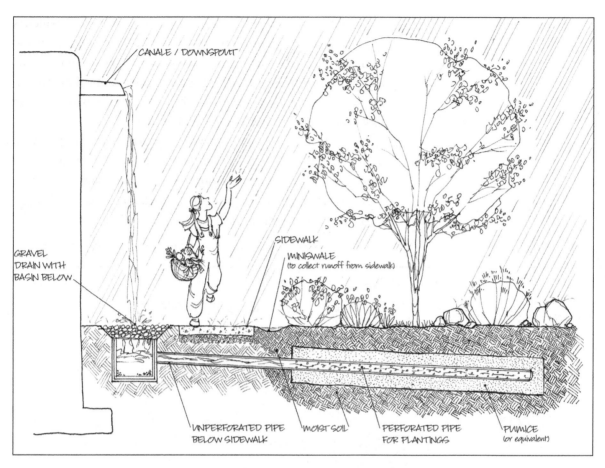

A Pumice Wick

You'll also get some great exercise digging a trench 8 to 10 inches wide and 16 to 18 inches deep. Assuming surface runoff is being properly routed, trenches should drop at least a half-inch every 10 feet—although development-review boards might want more of a pitch to your pumice-filled trenches, and the boards would be correct in insisting that wicks be installed at least 3 feet away from the foundations of any structures.

Depending on soil conditions, trenches longer than 20 feet should have perforated pipes running down the middle of the pumice so water will be released evenly. If you want to use water far from the *canale*, the location must be at least slightly downhill. Run 4-inch (nonperforated) pipe (at a one-quarter inch per foot drop) from the downspout or *canale* to the desired wick-start location, and begin pouring in pumice.

Depending on the depth of your trenches, add 6 to 14 inches of pumice to the bottom of your trench. Then, on top of the pumice add a layer of newspaper 12 to 24 sheets thick. This will prevent dirt from clogging the pumice. Backfill the trench with at least 3 or 4 inches of earth over the newspaper. Tamp down the backfill, and you're almost done.

The last step is for you to ensure that runoff has an appropriate place to go when the wick is full. To this end, simply make sure water coming out of your saturated wick at your box drain flows into a neighboring Santa Fe drain, French drain, check dam, or swale.

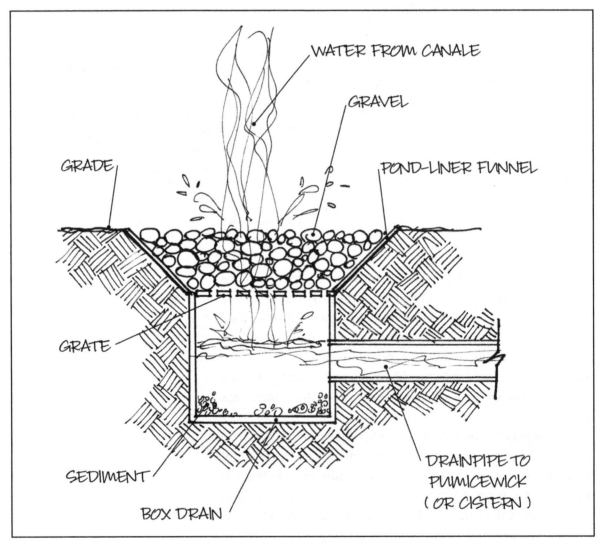

A Box Drain

Break the Wind

T. S. Eliot nails it squarely at the outset of "The Waste Land" when he says, "April is the cruelest month." A lot of people will tell you the famous ex-pat was referring to tax-paying season in the U.S., but reading further it's clear the poet was talking about northern New Mexico during our spring windy season:

What are the roots that clutch, what branches grow
Out of this stony rubbish?

A typically windy April in Santa Fe brings much the same thought. As dust, branches, and "stony rubbish" fly by you on the mesa, life gets pretty existential. Out-of-doors, grit fills the gaps between your teeth and gums, while unprotected eyes sting and burn, scratched by a stone haze blowing like a dry blizzard. Meanwhile, an incessant howl that's often easily heard indoors can become maddening.

These strong winds will, very effectively, remove moisture from any living thing—including healthy, well-mulched, and exquisitely graded soil. Your landscape's best protection is a well-designed windbreak of hardy tree and shrub species that can withstand and diffuse the wind around your more fragile plant material and your outdoor living spaces. In the case of cold north winds, you should also consider installing a windbreak around exposed walls of buildings that you wish to keep warm.

There are two types of windbreaks: impermeable windbreaks, e.g., solid walls and tight fences that deflect wind abruptly, and permeable windbreaks, e.g., plant material and spaced fencing that diffuse and gently decelerate high winds.

Solid windbreaks work to keep wind from small courtyards, porches, portals, doorways, entrance pathways, and tender plantings. Problems can arise, however, when the impermeable windbreak increases the wind's negative effects by diverting gusts around corners, into wind tunnels and sometimes into whirlwinds and little dust devils. Larger "devil winds" deliver their own brew of bad tidings as they scour down valleys and over windbreaks, but most winds can be diverted or deflected.

Permeable windbreaks come in two types of material: living and dead. Trees and shrubs constitute the live variety, while the lifeless kind refers primarily to widely spaced inert materials, such as milled wood or metal.

The two most serious "drawbacks" associated with living windbreaks are that they require significant time and need proper maintenance in order to successfully diffuse heavy prevailing winds or random storm events, whether wet or dry. Occasionally a temporary wood-fence fix will be replaced by a more permanent solution as trees come of age. Fortunately, within the gradual-greening system, we have plenty of time to care for our trees as they grow.

It's easy to imagine a windbreak translating into a much higher asking price in the local property market five, 10, or 30 years after its initial planting. And it's equally simple to envision happier households that are no longer burdened by high winds *and* taxes in April. These households just happen to have been owned by people with the foresight to plant a windbreak and a reasonable ability to keep trees and shrubs alive.

Like many water harvesting methods, establishing a planted windbreak requires water, but in the end a well-designed windbreak, like every form of passive water harvesting, helps store moisture in neighboring flora, fauna, soils, and mulches. If improperly designed, even permeable windbreaks can funnel forceful currents of air in at an unproductive angle, so be careful to plant your trees and shrubs in a subtle parabola whose tangent at its apex is perpendicular to the prevailing wind direction.

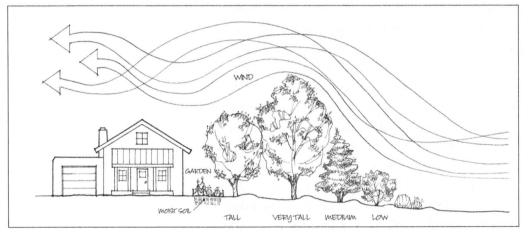

A Successful Windbreak

Avoid excessive gaps between trees, and make sure winds do not blow under the canopy of your trees toward people and plants. A row of shrubs can prevent wind from squeezing under the canopy of a mature windbreak, and a web of branching habits within the windbreak will often help fill gaps higher up the trees.

In order to cultivate a permeable windbreak, plant four zigzagging rows of plant material. Starting counter to the direction the prevailing wind is blowing, the first row should consist of a species that stays small when mature (such as shrubs or small trees suitable for coppicing and/ or pruning). The second row should consist of medium-sized species. The third row should consist of your tallest species, while the fourth row should have a mature height that reaches higher than the medium-sized species but lower than the tallest species.

The effect of designing a windbreak in this way is to direct wind up and over a house, a garden, or even an entire piece of property. This keeps moisture in the soil and, as an organic tool, passively harvests and maintains water and elements that would otherwise escape to the force of wind and evaporation. Consider plant species that are appropriate for your windbreak. For example, if you are primarily concerned with winter winds or spring winds, evergreens should be used because they will deflect wind more effectively than the bare branches of dormant or merely budding deciduous trees.

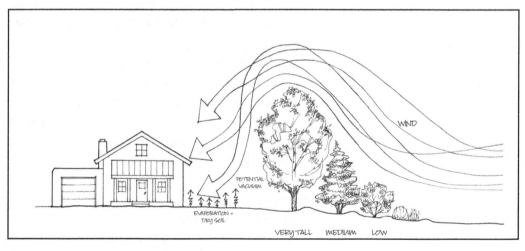

A Less-Successful Windbreak

Windbreaks provide shade, wildlife habitat, aesthetic beauty, privacy, noise abatement, firewood, and erosion control, but they should also be seen as landscape components that have a wonderful way of simultaneously increasing the value and elevating the comfort level of a piece of property. Most of all, any effective windbreak should be seen as a feature in a landscape that will reduce the need for supplemental water on the protected side of the break.

Create More Shade

Shade blocks the harsher impacts of the sun and the UV rays that prematurely age us and damage skin. Scientifically, it's the darker, cooler side of any object or objects that obstruct radiation from an energy source. Especially in semi-arid, arid, and hot climates, the draw of shade lures people, animals, and plant species alike. It doesn't take much to be aware of the comforting effects of shade that act to prevent the release of moisture from living bodies. Through a process called evapotranspiration, living things transpire and evaporate moisture at different rates depending on the intensity of heat and sunlight and the inherent characteristics of the species itself.

The creation of shade requires a kind of industriousness that resembles water harvesting more than water conservation.

Although we can think of shade as conserving moisture because it can reduce the need for supplemental water, to me the creation of shade requires a kind of industriousness that resembles water harvesting more than water conservation. This is not an essential point, but it's helpful from a gradual-greening perspective because it allows us to think of shade creation as yet another way of getting in that extra 10 minutes of gradual-greening time per day.

Many beneficial and edible plants that get the right amount of shade can actually do very well in water-deficient environments. Roses, for example, tend to get fewer diseases. The ancient agriculturalists of northern New Mexico grew squash between rows of corn in large part due to the substantial protection from the sun that the stalks provided. Shade, whether from trees, vines, shrubs, veggies, rock outcroppings, or built structures, is often the essential survival component for plants in an arid microclimate.

Also, shade can grow a property-owner's bank account or at least prevent money from flowing out of it as quickly.

Especially with rising energy costs and a worldwide need for trees to scrub CO_2 from the atmosphere, it is a wise decision to plant trees. Education courses and curricula at every level—from kindergarten to high school home-economics classes, as well as every architecture, landscape architecture, community planning, and business school on the planet—should teach about the positive effects of trees. Trees and other plant material can significantly reduce building/home energy requirements and in turn chip away at global warming by reducing and/or eliminating the need to run (or even to install) conventional, energy-gluttonous cooling systems. Air conditioners, swamp coolers, and electric fans typically use far more water than a well-chosen, properly placed shade tree.

Shade created by trees often increases one's quality of life at home, at work, and in your neighborhood. Take eating a meal. Given equally comfortable and convenient choices for dining, indoors vs. outdoors, most people will choose the outdoor option out of a natural instinct to feel simultaneously protected and free. It then becomes the work of the land steward, the landscape designer, or the gardener to provide as many niches as possible for people to enjoy their outdoor environment sheltered from the uncomfortable effects of too much sun.

In landscape design, creating shade is often about the development of "magic spots." These are places on any property where people can go have a quiet conversation or perhaps take time to read, reflect, stretch, and relax. Such places do not need to be completely secluded, and in fact a limited amount of communication among magic spots or between a magic spot and the landscape's main patio area is often desirable. In larger social situations, shaded areas provide an all-important getaway, private retreats where more personal conversations can take place away from the entire group.

On the financial side, magic spots increase the usable space of your property. Some realtors and assessors consider this a salient factor in assessing home value and saleability. Connecting the niches in your garden with pathways gives added form and purpose. Not everyone sees the value in such things consciously, but the comfort factor that hovers under any piece of shaded real estate is certainly improved when people feel protected from the elements and inspired by the connection to nature that these cocoonlike places suggest.

Although a magical spot next to a newly planted shade tree may take years to come about in full form, your gradual-greening investment will produce multiple returns adding up to more beauty, more comfort, more privacy, and much more value when it comes time to sell your property.

Leave Your Lavish Lawn Behind

The American lawn: Backyard barbeques and kids' swing sets. The perfect place for running bases. Dogs that jump up and catch a Frisbee. The symbol of our success rolled out like a red carpet under our sneakered feet.

Do I dare disturb this luscious universe? If so, how should I presume to begin such a vicious endeavor? It's no secret that the amount of water and chemicals we dump on the American lawn is so dangerous as to be nearly criminal. As an alternative, up until recently, I preached the virtues of plastic grass, which needs no watering, mowing, weeding, fertilizing, or spraying of any kind. Now, tests have found trace amounts of lead in New Jersey ball fields made of artificial grass—and the Environmental Protection Agency is "looking into it."

The manufacturers and distributors of the many brands of artificial grass/turf and related products are vouching for the safety of artificial grass even as their industry's trade groups are scrambling to do additional studies and assure the public that fake grass is safe and better than real grass. At this point I'm unsure. Additional questions are being raised concerning healthy levels of cadmium and phthalates found in some brands of fake grass, so all of this is making me take a second look at ersatz turf.

Several years ago we installed our own outdoor carpet under a mature shade tree. The "grass" was cut in a large circle around the tree and has suited its purpose very well, but now, for the fear that we might be putting our family at risk, my wife, Melissa, and I are rethinking its possible impacts.

Our fake sod certainly has a slightly scratchier feel than the normal wasteful, chemical-ridden lawn that covers so much of this country. It can get hot to the touch after several hours of intense sunlight. Arms and legs tend to break less with artificial grass according to a study of children's playing fields, but rug-burn counts go up whenever skin skids quickly on our glorified doormat. We'll have to monitor the situation with artificial grass as tests are conducted, regulatory agencies take a look, the industry reps give their talking points, and we assess what is safe and what is not. We of course need to greatly diminish the amount of chemicals used around home and property, whether on lawn or garden. Exposure to proven health hazards is the first step. Our families, kids, pets, playgrounds, and ball fields deserve no less than responsible awareness and action from educated adults and communities.

Fortunately, native plants take far less water to establish than a lawn, and native plants typically don't need supplemental water after they have been established. A lawn needs excessive amounts of water and pollutants and wasted time year after year. In contrast, by taking advantage of any of the techniques, systems, methods, and opportunities that I describe in this book, you can leave your lavish lawn behind. Something much better calls for you. Why not make it convenient, profitable, empowering, and enjoyable?

3

Active Water Harvesting Systems

But whosoever drinketh of the water that I shall
give him shall never thirst.
—John 4:14

Drink the Rain

At the turn of the 21st century, halfway through a wonderful honeymoon with Melissa "Down Under," we decided to visit a distant cousin living near Auckland, New Zealand. Over tea and sweet crackers, Rhoda, our hostess, discovered we were on an informal tour of permaculture demonstration sites. "Deary! Deeeary! Oh my! Oh, oh, oh my!" she exclaimed. "You *must* go and visit Paul!"

Rhoda sprang from her chair and waved her arms frantically as she dove for the phone. Within minutes, she'd organized a tour and dinner at her son's place a couple of hours south.

The next day, bright and crisp, we sped out of Auckland. The drive transitioned from a typical city to picturesque pastoral farmland that seemed familiar except for one detail: aboveground cisterns dotted the landscape alongside many of the homes, garages, toolsheds, and barns. Storefronts had cisterns almost encroaching into parking lots and sidewalks.

Eventually, we pulled up to a white metal gate, drove in, and found a nice flat place to park our camper van. Paul's dark green cistern was nestled between the garage and the main house. The tank was big and impressive, but it didn't overly dominate the lovely, cocoonlike transition from driveway to door.

We found Paul out back putting away tools. Soon we were touring the kitchen garden, the orchard, the pastures, and a small barn. We hopped into a lightweight Jeep to visit several neighbors with similar setups and then sped back to the spread for a quick shower, some fresh clothes, and table-setting chat. The meal consisted of almost only food that Paul and his family had grown within 1,000 feet of the dining room table—lamb, myriad steamed veggies, and a cornucopia salad that quickly brought out the best in "slow food."

During dinner we also enjoyed fine bioregional wine, fresh-baked bread made from locally harvested wheat and whole-grain milling, clear, cold water, and bountiful conversation with Paul, his wife, and kids, who were by far the most sustainable 21st-century people we'd ever met. At the end of the evening, as Paul walked us out to our car, we stopped and talked about his cistern system. "What kind of filtration do you use for your drinking water?" I asked.

"Filtration?" he asked quizzically, "Mostly, it's a piece of panty hose and a rubber band." For a moment I felt queasy and probably looked it because Paul chimed back quickly, "Same as all the neighbors!"

"And nobody gets sick?"

"Sick?" Paul chuckled, remembering something from long ago. "Nahw! It's just rainwater. Aint nothin' sick about it."

Recalling the heavy, sweating pitchers of water at dinner, I grinned. "Back in the states I know people who pump roof water through a bank of three different-sized micron filters and a pair of ultraviolet lights before they put it in their toilet bowls."

Paul beamed back in amazement and laughed. "Blooming Americans!"

Due to its independence and rugged isolation, New Zealand does pretty well on its own. It is a verdant land with some of the most spectacular scenery in the world. Fantasy buffs know it as Middle Earth because Hollywood chose to shoot *Lord of the Rings* in New Zealand, but few people know that it's also one of the early outposts of the international Green Party, which has now spread to over 100 countries. Sustainability—from water harvesting to full-on localized food production—is practiced in New Zealand as a cultural norm. If global markets ever completely collapse, it's hard to imagine any part of the modern world faring better than rural New Zealand. It would likely be among the most civilized and smoothly functioning societies on Earth.

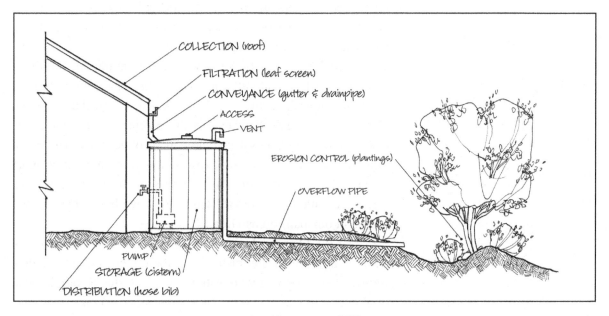

An Aboveground Cistern

We have pockets of people drinking their own harvested precipitation here in the United States, but it's safe to say that very close to zero percent of Americans use cistern water for anything, whether it be in the home or on the landscape. Even though cistern water is normally cleaner than well water and even though municipal water systems often poison their water with chemicals, habits are hard to change. My guess is that it'll be at least 10 years before a measurable percentage of Americans will be drinking cistern water, but that the movement will move much more quickly in the rainwater-for-landscapes sector of the cistern industry. Only as the terminology of active water harvesting finds advocates among community planners, architects, builders, landscapers, plumbers, and other professions will we see many permitting agencies start considering a roof as a source of household drinking water. It's my strong belief, however, that the next American revolution toward much more water harvesting will start *outside* the home with water being distributed on plant material. Winning the hearts and minds of the building industry will start in the areas around our homes—not at kitchen sinks or around the water cooler. Still, it's safe to say, once we have had proven success for a decade in our landscapes, the sink and the cooler won't be far behind because, ultimately, it's difficult to deny the fact that cistern water starts out cleaner than water that is sourced from a typical well or water-utility company.

In New Mexico, we have many examples of families, villages, ranches, and outlying communities that rely on cistern water for household use. On Rowe Mesa and in the towns of Cerrillos and Madrid, water, to the extent it's accessible via underground wells, lies deep below the ground often in a highly mineralized state. Cisterns provide answers to underground water that reeks of sulfur. Who wants to drink or shower in water that smells like passed gas? Don't we have enough foul smells in our bathrooms already?

My friends Jack Dant and Carrie McConaughy, who own a Santa Fe cistern-installation corporation called RainHarvest, are ones who walk their talk. Using a combination of aboveground and underground water-storage tanks, they live primarily on the minimal amount of moisture that falls on their home in nearby Canoncito. "We don't shower every day," McConaughy says with a hint of pride as she leads a local group of water harvesters around her home, "but when we do, we make sure all of that water goes out to our fruit trees down by the trampoline. We don't water these trees with anything else." When asked about filtration, she replies with confidence, "We haul in our drinking water, and we have a well that produces an extremely low trickle. The rest of our water comes from the sky. We run our harvest through a series of screens, baskets, and sediment filters, and as long as we keep the tanks clean, we're good to go."

Toward the end of the tour, Dant wraps up their cistern-oriented perspective honestly, "We've learned to live within our means. And we have a very real appreciation for the scarcity of water—both coming from the ground and from the sky. We have learned not to waste water, knowing that as a result our high-arid garden will benefit."

As global climate change and regional meteorological shifts threaten harsh conditions and droughts, more people will be making the switch to new water and energy solutions. We can expect smart homeowners to set up their homes to take advantage of all alternatives including fully plumbing homes to collect, convey, store, and distribute potable rainwater. Readers interested in the best system for ensuring that your roof runoff is safe, please read the second section of chapter 4, "Wastewater Harvesting Methods."

For the "blooming American" in any of us, the official "how to" part about drinking precipitation involves wastewater issues because of the potential for accidental contamination from animals that could climb into the cistern and drown, the slim chance for bird and squirrel poop to create dangerous bacterial or chloroform problems in your tank, and a more significant potential for human error one day causing pollution in a cistern. Rainwater is not wastewater, but when it comes to promoting the drinking of rainwater to a culture that is used to drinking a soup of chlorine, fluoride, and any number of pharmaceutical products, we must tread with special care and attention.

At the moment, we have some significant work to do in order to bring a cistern-illiterate society up to speed. The best way to do this, I believe, is gradually through water literacy, training in landscape design, and a thorough study of everything from the microclimates that exist on a piece of property to a brainstorm of all of the potential uses a parcel of land might have. Early on in the landscape-design process, all of the challenges with regard to your water systems, including a possible cistern (with its size, shape, placement on the site, relationship to grade, quirks of the particular materials chosen, installation, and maintenance), need to be addressed. If cisterns are poorly placed, installed improperly, and are difficult to maintain, then cisterns will get a bad reputation, so this must be avoided at the outset of the upcoming water harvesting revolution, and for now, in order to grow gradually green let's start where using cistern water is safest and easiest, that is, in the landscapes we create.

Create Jobs

Cisterns come in almost any size or shape including spheres, cylinders, cubes, and all lengths of sausage-shaped tanks. They can be installed underground, aboveground, or partially buried. From concrete to plastic, cisterns can be made out of at least eight different materials. In the aboveground and partially buried tank scenarios, dark-hued tanks work best because they will be ultraviolet-light resistant.

Every tank needs an air vent, an overflow pipe, a lockable serviceway (i.e., a manhole), a delivery point, and an exit point into a main distribution line. Most storage systems require pumping, but they can be gravity powered if your roof and associated water usages are perfectly situated on a steep enough slope. Options include float switches to prevent pump burnout, extensive filtration, ladders or a rope for climbing in and out of the tank, and alarms that sound when leaks are detected.

These and many more decisions must be made by whomever is designing the system. For example, some systems include automatic valves that import water from auxiliary sources (like wells, reservoirs, rivers, etc.) during dry times. These systems provide a level of convenience that many people desire and expect. Less sophisticated systems usually have a backup hose for the same purpose of adding a little water to the system when it's low.

In the following chapters you'll discover helpful details about all of the above. During the remainder of this short section, we'll address the extremely important issue of cost as we simultaneously see how water harvesting might become an important part of our nation's economic recovery. Through it all, please keep this thought simmering somewhere: Every cistern system signifies numerous green-collar jobs.

Cisterns can be expensive due to the many components associated with the design, planning, surveying, materials, organization, excavation, pipe fitting, tank installation, electric-line connections, and, finally, all of the miscellaneous vents, overflow pipes, float switches, pumps, pump houses, plumbing components, irrigation valves, hose bibs, rain sensors, and so on. Add hands-on labor, equipment rental, delivery, permitting, insurance, overhead, and debt service. Meanwhile, the costs of regular maintenance and sporadic repair need to figure into any cistern-cost equation.

I'm in the business of designing ecological landscapes, which means that many of my clients want cistern systems integrated into their designs. This is almost always possible to accomplish in theory, but often the project can be impossible to accomplish for reasons of simple economics.

In the past, most people wanted underground, automated, low-maintenance systems. Depending on the size of the cistern, existing site conditions, and a variety of other factors, these systems typically range between $20,000 and $40,000. These days, small-scale, aboveground, gravity-fed systems that require more hands-on ability and fewer upfront costs are becoming increasingly popular. Ideally, aboveground or partially buried systems on sloped terrain can provide a naturally pressurized valve at the bottom of the cistern. This kind of system can sometimes be installed for between several hundred and a few thousand dollars.

If the U.S. government wanted water independence, considering these costs, we could install a cistern for every household in the United States for around the estimated $2 to $3 trillion cost of the war in Iraq. The bottom-line question is infrastructure spending versus wasteful spending, and, unfortunately, the 43rd U.S. president with congressional approval supported the worst form of waste: unnecessary violence. Fortunately, looking at our current economic situation, with the collapse of cheap money, the implosion of the real estate bubble, and the pervasive environment of lower consumption, at least the possibility of profound positive change finally exists.

Active rainwater harvesting needs to be offered as an option in every construction-spending decision at every level of government.

As the Obama administration ramps up the largest building programs since Roosevelt's response to the Great Depression, a jump start is in the cards in which energy-conscious and environmentally forward-looking projects are coming into their own. Now is the time to make water a critical part of this mix. Active rainwater harvesting needs to be offered as an option in every construction-spending decision at every level of government. In addition to rebuilding our nation's traditional infrastructure, why not improve it by turning blue-collar workers in construction and all related fields into the green-collar workers that Van Jones describes in *The Green Collar Economy*? By putting people to work on water harvesting projects, skills will be learned while money is earned and a tangible result with multiple benefits will be attained. Begin with existing projects in the pipeline for federal funding and annotate requirements to add sustainable-design elements to all. Offer direct funding to communities that meet Leadership in Energy and Environmental Design (LEED) standards for water harvesting and watch, over the long haul, the values of these properties rise faster than those of neighboring properties left behind in the development strategies born of the last century.

After a few years of direct federal revenue grants and funding, the government could pay half of water harvesting's costs and homeowners or developers could pay the second half. A few years after that, there should be enough interest in water harvesting that subsidies would not be needed because the virtues of water harvesting would be well known. Costs will have come down and local industries will be established, working with energy sufficiency, wind, and related alternative technologies to create a nexus of the next generation of our economy and our leadership role in the world.

In some communities, government buildings might set the scene for training sessions for up-and-coming water-conscious professionals. Active and passive installations will provide hands-on experience for an army of water harvesters who will be setting in place skills and talent we might need to address upcoming water shortages. At the same time, an army of tree planters could be scheduled to come behind the water-system installers in order to plant shade trees that in turn reduce the air-conditioning costs of the same government buildings.

Already, green developers are realizing that shared water-saving systems can be very cost-effective. The pumps, tanks, backup water connections, pressure tanks, and many more costs will bring economies of scale if planned and accomplished in aggregate. When costs are spread out over 10 or 12 homes, or a number of government buildings, significant money can be saved. The work ahead will not stop at water harvesting. As a foundation of sustainability, however, it seems a proper place to start. The long-term water supply should be the highest priority of a forward-thinking society, but because as a financial system we tend toward short-sighted quarterly returns as we look at investments, this doesn't mean that we should overlook the steady, gradual returns of smart water harvesting. Water, energy, and a sustainable mix of permacultural solutions is truly the ticket to success over time. Harvesting our natural resources with a wise, considered approach is the key to tomorrow's economy.

Examine Cistern-System Anatomy

Welcome to the heart of the matter. Now that we've established the potential in terms of the resources and economic stimulus that cisterns can bring to our lives and our communities, let's look at the anatomy of these systems that collect, convey, filter, store, and distribute precipitation by means of a tank called a cistern. In order to complete the harvesting process, stored water is sent to an intended location on a given property. Harvested water is usually filtered both before and after storage. Think of a cistern as an extremely convenient miniaquifer or a highly localized reservoir.

The installation of a cistern is a perfect example of how it is sometimes best to bank those 10 minutes of daily gradual-greening time into a longer chunk of time in your schedule. The design and installation of an efficient and productive rainwater catchment system takes a significant commitment of time and energy.

Even if you are having a professional install and maintain your cistern system, you will still want to put some time into learning about the basic components of it, how storage systems function, what to look out for, and what to take advantage of. Whether you happen to be a do-it-yourselfer type or a call-the-professional type, this part of the book is designed for you.

Cistern systems usually require more planning and more financial resources than passive water harvesting systems, but the basic concepts involved are still easy to grasp. While passive systems require no moving parts and very little maintenance, active systems often include parts that need regular upkeep. Such systems always feature a storage tank to allow water to be held and later distributed on demand. Whether or not you can get away with a system as simple as Paul's in New Zealand, your active water harvesting system should be comprised of the following:

Collection: Surfaces that capture and redirect precipitation

Conveyance: Conduits capable of diverting collected precipitation toward a cistern

Pre-cistern filtration: Contraptions attached to the

conveyance component that prevent debris from entering a cistern

Cistern: Water storage tanks or similar vessels that hold water

Overflow: Pipe that allows excess water to exit a full cistern

Air vent: Conduit or orifice that allows air in and out of a cistern

Post-cistern filtration: Filtering unit associated with the hydraulic lift and distribution components that prevent undesirable solids from passing through the system

Hydraulic lift: Any means of intentionally removing water from a cistern

Distribution: Any conduit or fixture that delivers water to an intended place

As a device for helping us think about these components, consider the anatomy of a human circulatory system. The collection surfaces of a cistern system resemble the capillaries at the outer branches of our veins, while the conveyance system resembles the venous tributaries that make up the side of the human heart where blood enters. Methods of pre-storage filtration mimic the pulmonary system, which cleanses the blood of carbon dioxide.

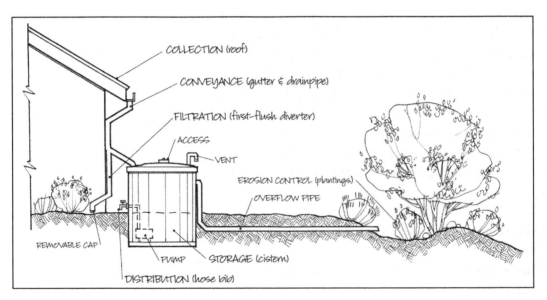

A Partially Buried Cistern

Even as cistern systems resemble a circulatory system, the tanks themselves tend to be reminiscent of the human bladder, whose critical function is the storage of liquid. Also like a bladder, cisterns can get full, and when they are full, they need a conduit for overflow, that is, an exit strategy.

However, since many cistern systems have submersible (sump) pumps installed directly within them, the cardiopulmonary analogy is ultimately better than the urinary analogy, as one considers how the lungs resemble a cistern's venting and filtration components and how a cistern's distribution system will have major and minor arterials, just as a mammalian circulatory system has large and small arteries. Plus, whether these arteries are pressurized pipes, shower heads, or drip irrigation tubing, they often resemble the branching patterns found in arteries.

A cistern's purpose, like that of a heart, is to keep a system alive. Its shape, like that of a circulatory system, consists of two sets of branching conduits with a central vessel connecting them in the middle. Its function, like the pumping nature of any cardiopulmonary system, is to move fluids from place to place. And its basic physical substance is as vital to life as is blood. "Water is life" has a resonance we can feel as we collect and distribute the water that keeps our systems alive.

Catch the Rain

Roofs, patios, pathways, sidewalks, parking areas, roads, and even natural slopes can all be collection surfaces, the capillaries of an active water harvesting system. As a water supply, roof water is typically cleaner than water harvested from other surfaces. Meanwhile, since roofs are elevated above the ground, they can provide a significant amount of hydraulic lift. For these reasons, roofs are typically the catchment surfaces of choice for cistern systems.

Patios, pathways, sidewalks, and other pedestrian-oriented surfaces can be easily used to collect precipitation. The drawback is that these surfaces tend to create more cleaning and maintenance as they increase the system's need for filtration both on the conveyance side and the distribution side of the cistern. In New Mexico and many other states, it is even illegal to harvest nonroof water into cisterns due to interstate legal compacts mandating surface flows to downstream water users.

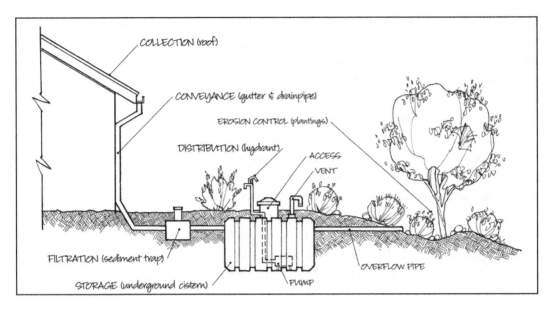

An Underground Cistern

When driveways, roads, parking areas, and other vehicle-oriented surfaces are chosen as collection surfaces, special care must be taken to remove pollutants before, during, and after water storage. Similarly, water collected from natural slopes will contain significantly increased levels of particulates. Filtering suspended particles creates more expense for both the installation and the maintenance of your system. For these and other reasons, water collected off of roads and slopes is typically, on a residential scale, best diverted to passive systems such as swales, French drains, Santa Fe drains, pumice wicks, buried *ollas*, and many of the other techniques described in chapter 2.

On a larger scale, a number of filters can easily, though not inexpensively, produce clean water out of stormwater runoff using a wide variety of techniques. Using standard screening and filtration techniques along with volumetric-, hydrodynamic-, and oil/water-separation techniques, even the dirtiest of parking-lot grime can be cleaned to any desired level. The Vortechnics line of filters available through Contech Construction Products would be where I would start research in this regard.

From a collection standpoint, the most important architectural question concerning your roof will be whether it is pitched or "flat." For water harvesting purposes, pitched roofs are preferred, but they are by no means required. The next section, "Choose a Pitch," explains why roofs that have a significant pitch are better at harvesting water than roofs that are only slightly sloping (as is typically the case for those that are said to be flat).

Another collection question that comes up for home builders revolves around the type of material that should be used as a given cistern system's collection surface. The upcoming section "Consider the Material" provides tips on appropriate materials ranging from metal and slate to rubber and tar.

Finally, the size of your collection area often can be greatly increased if you think outside the box in terms of the roof surfaces that may be available to you. This is to say that the roof of your house is often just the beginning when it comes to available roof collection surfaces. Don't forget to harvest precipitation off of garages, carports, porches, *portals*, toolsheds, and other outbuildings. Your net harvest can be substantially increased in this way as "Size Up Your Roof" will explain.

Choose a Pitch

You don't need a crystal ball to see a shape-shifting future for architecture wherever water is scarce. For example, New Mexico's traditional flat roofs will gradually fade into history as people become more interested in harvesting the rain and snow that fall on their roofs. It's not that flat roofs are terrible for water harvesting; it's just that pitched roofs are much better.

First, when precipitation lands on a pitched roof, water immediately slides down the slope of the roof. By contrast, when rain, snow, sleet, and hail hit flat roofs, a substantial amount of your potential harvest is lost to evaporation while the precipitation puddles and sits on the roof for a period of time. Much of this moisture makes it off the roof, but a significant portion (estimates range from 10 to 50 percent depending on roof material and type of storm event) dries up and disappears.

Second, when rainwater and snowmelt flow into the gutters and downspouts associated with pitched roofs, they can be easily harvested at an exact location. When runoff pours sporadically out of flat-roof outlets, *canales* as they are called in New Mexico, it is usually much more difficult to collect the water due to the wide variety of places where the potential harvest might land.

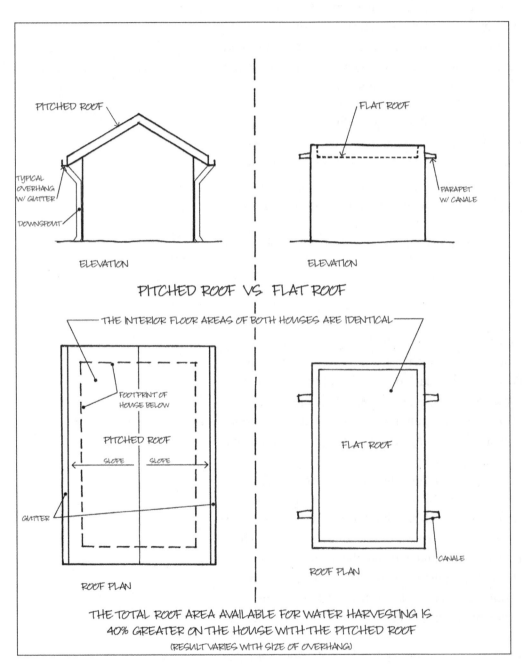

PITCHED ROOF VS FLAT ROOF

THE INTERIOR FLOOR AREAS OF BOTH HOUSES ARE IDENTICAL

THE TOTAL ROOF AREA AVAILABLE FOR WATER HARVESTING IS
40% GREATER ON THE HOUSE WITH THE PITCHED ROOF
(RESULT VARIES WITH SIZE OF OVERHANG)

Pitched vs. Flat

Third, when you convey roof water to an appropriate place in the landscape via the gutters and downspouts of pitched roofs, it is far less expensive than the underground conveyance work involved with flat-roof runoff. Digging trenches, installing and fitting pipes, backfilling and tamping trenches, and all of the other work described in the sections that follow about conveyance methods are all time-consuming and costly activities that can be minimized or completely avoided with gutters and downspouts. You can attach aboveground pipes to *canales* or even in-wall conveyance pipes to flat roofs, but this should be done only after weighing all the aesthetic, structural, and financial factors involved.

Fourth, a cistern system's overall installation cost will usually be significantly less when water is collected from a pitched roof. The main reason for this is that more *canales* are required for a flat roof than downspouts are needed for a pitched roof. Building codes often require at least twice as many *canales* as downspouts per square foot of roof area.

Fifth, flat roofs are harder to keep clean than pitched roofs, gutters, and downspouts. In general, pitched roofs provide a cleaner water supply because less debris accumulates on pitched roofs than on flat ones, while flat roofs typically have parapets that prevent debris from sliding or blowing off of them, so when significant precipitation events occur, whatever still remains on a pitched roof will quickly be washed away. Conversely, at least some portion of the debris on a flat roof is left behind to decompose on your water-collection surface.

Sixth, many materials used in flat roofs leach low levels of toxic chemicals. If you are harvesting precipitation for household purposes, these toxins can be eliminated with filtration. If you are simply harvesting water for use in your landscape, typically your plant material will not be affected significantly by the toxins. If your filtration system is relatively primitive (which it can be for plant material, toilet flushing, and car washing), marking your faucets and valves as "nonpotable" and coloring them with purple paint to alert potential users is required in many jurisdictions—regardless of the angle of your collection surface.

Keep in mind that, ultimately, it doesn't matter what kind of roof you have. Although there are many reasons to build pitched roofs rather than flat ones, none of these reasons are

Ultimately, it doesn't matter what kind of roof you have.

extremely significant. What's most important is that you make an attempt to harvest the rain from whatever type of roof you have. Use what you have as a starting point and continue your own personal process of gradual greening from that unique and potentially productive place.

Consider the Material

Another collection question hovers over roofing materials. The easiest way to describe these materials is to maintain the distinction between flat and pitched roofs. An extremely wide variety of roofing materials is used all over the planet, so we will not attempt to discuss every type here. But, in the two sections that follow, you will discover a number of important characteristics associated with some common roofing materials.

Before we get to this, it's worth mentioning that no matter what kind of structures you have, from cathedral to grass shack or from the Federal Emergency Management Agency headquarters building in Washington, D.C., to a one-room disaster-relief trailer, you can harvest precipitation off of its roof. The one exception would be if your roofing materials happen to be toxic. Fortunately, the use of lead paint and asbestos roofing has declined significantly since 1980, but trace amounts of asbestos can still be found in some asphalt shingles, asphalt roofing felt, and cement roofing shingles. Avoid roofs filled with asbestos and walls coated with layers of old lead paint when purchasing a house. Serious health-related problems often occur during the removal of either, so if you happen to live in an asbestos or lead-laden house, you may want to move or be sure to carefully vet and supervise the licensed professional you choose to remove said toxins. (Never commence a do-it-yourself sanding project to remove lead paint from walls/window sills/banisters, etc. when children are present. Even minimal amounts of inhaled lead dust can have long-term neurological consequences.)

For good reason, lead, which is great for soldering metals together, is not a roof material per se, but it is sometimes used along the seams of canales, downspouts, and skylights. My wife and I know firsthand because we discovered trace amounts of lead in our cistern water. Fortunately, we traced the source back to the soldered seams between and among the cut-metal surfaces on our canales. When you add up the biking to Empire Builders, the purchasing of a roof-patching tar, and the spreading of the sticky stuff all over the lead seams, preventing serious lead pollution in our cistern water only took about a week's worth of gradual-greening time. It was time well spent and time that I will want to spend every year, like Robert Frost mending his wall in spring, in order to make sure that the lead seams always remain completely covered by my less-toxic patches.

It's worth noting that with subsequent soil tests taken from the soils that we watered most often, we now know that the lead threat was nonexistent on our property. All of our 15 soil samples and the chicken eggs (we let our chickens consume cistern water, but we prefer other sources for our own drinking) fell way, way below any dangerous level and very far below the average lead content for soils in the United States. Still we sleep better at night knowing that the lead on our roof is covered by a common roof-seaming material, **tar.**

The most common pitched-roof materials are described below in alphabetical order:

Asphalt shingles are popular because they are relatively inexpensive. Since they are made from materials marinated in asphalt, know that some leaching will occur. Even though common roofing shingles are not the best roof choice from a water-quality perspective, asphalt particulate can be removed with filtration in potable-water-use situations, while intense filtering can be ignored in non-edible-landscape situations.

In some parts of the world, **clay tiles** are the dominant roofing material. Unlike asphalt, clay is smooth and, as such, it is highly efficient for water harvesting. In many places tile is not especially popular because of the increased materials and maintenance costs associated with it. Clay should be avoided if significant freeze-thaw conditions occur in your area. Such conditions can create crumby tiles, which not only can cause costly roof problems, but also create more work during the filtration processes.

Laminated fiberglass shingles contain the same materials as some asphalt shingles, but they last up to twice as long as regular asphalt shingles. The potential leaching of asphalt and fiberglass make this a poor choice from a water harvesting perspective, though.

For the money, a pitched, **corrugated metal** roof is usually the cleanest, most efficient, and most cost-effective of all precipitation-collection surfaces. Metal can also last a relatively long time. It is neither the most expensive nor the least expensive roofing material available, but the bottom line is that, compared to every other type of roof, metal loses the least amount of precipitation to evaporation and is second only to slate from a water-quality perspective.

Slate is an extremely durable collection surface. Like metal, it usually provides very clean water. It typically comes in the form of small rectangular tiles, often called shingles. It can be very expensive in part due to shipping costs from quarries in the northeastern part of the United States.

Wood roofs are usually built using shingles made either of a hard wood that contains a natural resistance to rot or of wood that has been treated with a rot-resistant chemical. Wood roofs are often very attractive, but they are usually relatively expensive and are typically the most flammable of standard roofing material available. Also, the leaching of rot-resistant chemical particulate may negatively affect the water quality in your cistern.

As discussed above, so-called flat roofs are not the preferred style of roof from a water harvesting perspective, but they are reasonably effective at harvesting precipitation nonetheless. When the collected water is designated for nonpotable uses, such as plant material, the drawbacks of flat roofs, from both the water-quality and -quantity sides of the question, are often relatively minor. So, if your roof happens to be flat, do not despair because you can create a totally aquifer-independent landscape with any of the following common flat-roof materials:

Due to its relatively low cost, a wide variety of **asphalt-based** roofing materials for flat roofs are common throughout the Southwest and beyond. Unfortunately, all asphalt-based products will leach toxins, so tar-and-gravel and other asphalt-based roofs should be avoided in situations where potable water is required. Modified bitumen is an asphalt-based product containing a modifier that increases its waterproofing characteristics. These modifiers include rubber, polyester, fiberglass, and atactic polypropylene. If properly installed, such roofs should last longer than typical tar-and-gravel roofs, but they also require more skill to install than most flat asphalt surfaces.

Another alternative is **polyurethane foam.** It leaches fewer toxins than asphalt, but it is slightly more expensive. Polyurethane foam is an excellent insulator but should be covered with a UV-resistant coating, or, as is the case for many asphalt roofs, the surface must be covered with a thin gravel layer.

Rubber, or EPDM rubber (ethylene propylene diene monomer), is becoming an increasingly attractive flat-roof material. It is priced competitively and is long lasting and durable. Rubber is less toxic than asphalt, requires relatively little maintenance, and holds heat within a home rather well.

Size Up Your Roof

In addition to considering the angle of your roof and material of which it is made, you should also pay attention to the size of your all-important collection surface. In rainwater harvesting, size matters a great deal.

Forever desirous of being on the cusp of all things sustainable, but more importantly wanting to beat the City of Santa Fe to the punch (in their not always unspoken but generally healthy competition), a few years ago Santa Fe County came up with a forward-thinking building ordinance that requires the installation of cisterns for residential buildings over a certain size.

One peculiar size-related issue came up just as the county commission was about to vote in favor of the ordinance. At the 11th hour, the language in the resolution requiring the installation of a cistern was amended to read "2,500 square feet of heated area" rather than "3,000 square feet of roofed area." Someone realized that the "roofed area" language might discourage the installation of large portals, which represent a relatively inexpensive way to enjoy the great outdoors— under an extended portico. The problem is this change could encourage people to build three-and four-car garages that they quickly convert into bedrooms once the inspectors have come and gone. Ultimately, however, this is a problem worth living with because roofs are inexpensive compared to heated space and, in the world of active water harvesting, governments should encourage (rather than discourage) the construction of large collection areas.

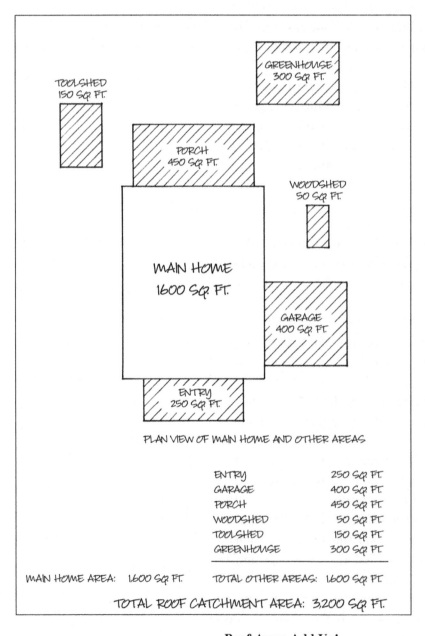

PLAN VIEW OF MAIN HOME AND OTHER AREAS

ENTRY	250 SQ. FT.
GARAGE	400 SQ. FT.
PORCH	450 SQ. FT.
WOODSHED	50 SQ. FT.
TOOLSHED	150 SQ. FT.
GREENHOUSE	300 SQ. FT.

MAIN HOME AREA: 1,600 SQ. FT.   TOTAL OTHER AREAS: 1,600 SQ. FT.

TOTAL ROOF CATCHMENT AREA: 3,200 SQ. FT.

Roof Areas Add Up!

Here's why: Determine the size of your roof in square feet by multiplying the length of the roof by its width. Note that the increased elevation of a pitched roof does not increase your catchment area by virtue of its steep incline. While it is true that more materials are needed to cover a pitched roof, the same square footage of property is covered for pitched and flat roofs of the same dimension. Each roof still covers the same amount of surface area as a flat roof of the same length and width.

One size-oriented advantage that pitched-roof structures have over flat-roof structures is that pitched roofs often have large overhangs. This means that given the same exterior wall dimensions, a pitched-roof house will have larger roof dimensions than a flat-roofed house. The effect of a 2-foot overhang all the way around a pitched-roof 5-foot-by-10-foot toolshed is a 160 percent increase in square footage over a flatroof covering the same amount of indoor space.

Meanwhile, in your calculations of square footage, remember to include every other kind of outbuilding or breezeway imaginable. These are valuable resources that should be harvested, keeping in mind the gradual-greening system and adding additional structures to your total collection area as more of your time and money become available.

Dare to Be Normal

Now that you have determined the square footage of your collection area, it's time to estimate the amount of precipitation you can expect to get each year. An easy process, *que no*? We think that calculating our normal annual harvest should simply be a matter of multiplying the total square footage of collection surface(s) by your location's average annual precipitation rate.

The problem is that, especially in many arid lands, *it is abnormal to be average*. In fact, in a "normal" year in the American Southwest and northern Mexico, we get 20 percent less precipitation than we do during an "average" year. This is due to those rare years when rainfall is atypically heavy, which skews the average up in such a way as to be misleading when what you really want to know is how much water you might expect from your roof.

In order to determine the size of some of the key

components of your water harvesting system (including conveyance piping, size of your cistern, and size of your cistern's overflow pipe), it is essential to make an educated guess as to the amount of precipitation you are likely to harvest in a normal year and extreme years alike. You will want to check to determine whether there's a difference between normal and average years in your bioregion. In many moist, coastal areas, there's no significant difference between them. Wherever you live, when sizing cistern-system components, you will also want to be prepared for any extreme flood event.

Once you have established how much rain you're likely to get *in a normal year*, multiply this quantity by the square footage of your total collection area. In Santa Fe, for example, where average precipitation is 12 inches per year, normal rainfall happens to be 10 inches per year, so one square foot of roof will provide .83 feet of water in a normal year. Multiplying this number by the number of gallons in a square foot (7.5), we discover that 1 square foot of roof gives us 6.2 gallons in a normal year.

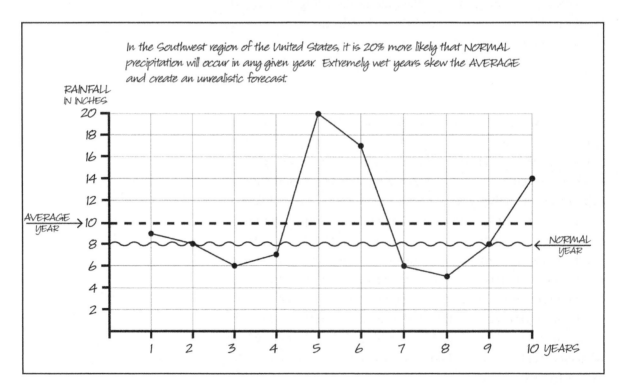

Two Views of Annual Precipitation

This means that a 1,200-square-foot house with a 200-square-foot covered back patio, a 100 square-foot front portal, a 100-square-foot toolshed, and a 400-square-foot garage will catch over 12,000 gallons of water in a normal year. Not bad for an arid climate and plenty of water to establish a drought-resistant landscape. But more importantly, in the gradual-greening scheme of things, once windbreaks are well established you can soon begin to provide for a good-sized vegetable garden complete with perennial fruit-bearing bushes and trees. If you add greywater recycling, on-site sewage treatment, a little community water harvesting, and some of the passive water harvesting techniques described in chapter 2 into the mix, full blown sustainability is ultimately possible.

Convey Your Dividend

Conveyance piping diverts collected precipitation to an entry point in the cistern. These veinlike conduits can be dissected into three essential parts, described in detail below. A fourth component that is typically found within conveyance piping, pre-cistern filtration (see next chapter), does little in the way of moving runoff toward your tank, but it is very important because it acts to clean collected water en route to the cistern.

Just as it is difficult to imagine a circulatory system without veins, it is hard to visualize a cistern system without a conveyance component, and just as veins can get extremely complex, so too can those related to cisterns. Because roof runoff can be seen as a kind of dividend that pays you in the most necessary currency for life on Earth, it's important to understand the three basic elements of this critical cistern-system component:

> **Focal points:** locations, such as gutters or *canales*, where roof water is concentrated before proceeding into the system's drop lines or lateral pipes
>
> **Drop lines:** places, such as a downspout or the space under a *canale*, through which collected precipitation drops from the roof into the lateral pipes of the system or into the cistern itself

Lateral pipes: sideways conduits that direct roof water either from a focal point or drop line to the cistern

Focal points are the first veins of what I call the "conveyance stream." In pitched-roof scenarios, the focal points are the gutters that channel roof water toward downspouts. Gutters typically are made of either galvanized steel or aluminum, but copper and stainless steel gutters can be used if you want to spend extra money for some aesthetic reason. They should drop about 1 inch for every 40 feet of run, and they are typically at least 5 inches wide and wider in steep-roof and large-roof situations.

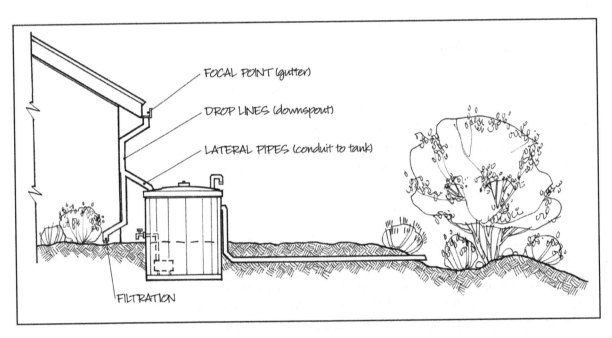

Conveyance System for a Pitched Roof

Gutters require periodic maintenance: cleaning out leaf matter, pine needles, bird droppings, dirt, and other garbage. (This gunk and debris can usually be tossed directly into the compost pile.) If you are fortunate enough to have tall trees growing near your house, your good fortune will include having to do more gutter work compared to people who have no mature trees, branches, birds, and squirrels up over the roof or in the general vicinity. Keep in mind that this chore usually takes much less time than all of the procrastinating associated with it!

Focal points for flat roofs often look like modernist gargoyles protruding out of the top stories of buildings. In the Southwest these *canales* are common in pueblo-style architecture. In cold climates, they should be sited on the south- and west-facing sides of a house to avoid the formation of icicles and blockage. If a north-side *canale* is blocked by anything during severe winter weather, enormous stalactitic ice appendages can grow out of your house and wrench your *canales* from walls as they slowly heave toward the ground.

Builders should install *canales* as close as possible to the chosen site of the cistern. First, this will save money on installation costs by keeping lateral pipes short. And second, unnecessarily long lateral runs can create annoying underground obstacles in your landscape when you come back to plant a tree or build a deck and find that a just-broken conveyance pipe is in the way.

Flat roofs sometimes incorporate drains directly within the walls of a building. The drawback of such in-wall pipes is that serious problems would likely occur if the pipes ever leak. But when one considers the fact that modern people have had pressurized water pipes hiding within the walls of their houses, offices, factories, schools, and stores for multiple generations, it's easy to have faith that our engineers, architects, and builders will be able to prevent pipes with water at a lower pressure from causing much damage.

Three complications, however, exist for drain pipes that are installed within the walls of buildings. One is that, unlike the leak in a pressurized pipe, a small leak in an in-wall pipe could go undetected for a long period of time. Further, it is easy to imagine such a leak occurring far away from where it is initially detected. Finally, when a leak in a pressurized pipe occurs, you can shut off the water during the time when you want to fix the leak. You can't stop the rain, and it is often difficult or dangerous to prevent water from using its designated conveyance piping because other conveyance pipes on the roof (if they are accessible) will probably not be sized large enough to handle an extra drainage area during a large storm event.

Just as other branching systems tend to have sharp

angles, conveyance systems almost always feature at least one place that has a steep drop line at a right angle to the movement of water at a focal point. Downspouts connected to gutters tightly control this water in a conduit running down the walls of the structures associated with them. Downspouts are typically made of vinyl, PVC, wood, galvanized steel, corrugated metal, aluminum, and other metals. In order to prevent leaks, downspouts should be fabricated out of the same material as your gutters because leaks often occur at the seams between different materials. They are normally rectilinear, and their dimensions are usually either 2 inches by 3 inches or 3 inches by 4 inches, but they can come in cylindrical shapes, too.

Unless your gutters already exist and are made out of some other material, for the do-it-yourselfer, vinyl is an excellent downspout material. It comes in manageable sizes and parts that fit together easily. Professionals often use vinyl gutters and downspouts because the material costs are less than for other gutter materials and because vinyl is so easy to install.

PVC provides a smooth, clean surface that can connect easily with horizontal conveyance piping. Unfortunately, as mentioned earlier, PVC carries a set of risks related to producing toxic dioxins during its production and eventual disposal, so from an environmental perspective it has some serious drawbacks. Its popularity is due to its relatively low materials cost combined with its ease of installation.

The best way to capture the downward diversion that pours off *canales* is with a drain that is slightly below the surface of the earth a few feet out from the base of an exterior wall. For a variety of reasons ranging from the aesthetic to the functional, a top dressing of pebbles, gravel, cobbles, or rocks is often used to cover the drain box and the plastic funnel associated with it.

Cistern systems usually need lateral conveyance pipes connecting the drop lines to the cistern. Since they must ensure positive drainage toward the cistern, in some situations pipes can drop at a rate of 1/8 inch per foot, but 1/4 inch per foot is typically required by building inspectors.

Lateral conveyance pipes can run high aboveground, underground, or on the surface of the ground. Often, elevated lateral piping will direct water toward an aboveground cistern, while underground laterals will connect to subsurface tanks. Do-it-yourself cistern-system designers can get in many of their gradual-greening hours in any given week working on everything from gutter-materials research to city regulations on how far away underground lateral pipes must be from a given structure, such as a home or garden wall.

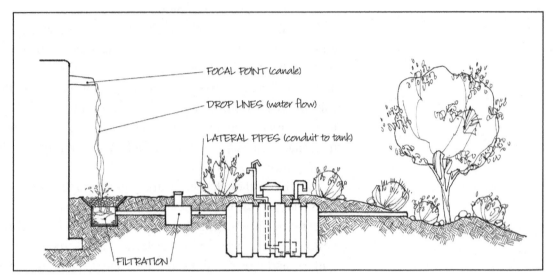

Conveyance System for a Flat Roof

In order to ensure that the investment you make in a cistern system reaches its potential, it is important to prevent damage to your home. Designing, installing, and maintaining all of your conveyance piping is a task to take seriously. Pipes should be sized according to local requirements, but most residential applications can use conveyance piping in the 4-inch -diameter range.

Filter the Rain

Filtering the collected precipitation before it ends up in your cistern is one of the easiest ways to keep harvested water clean. Whether you plan to use your harvest for drinking or washing your car, landscaping, or clothes washing, pre-cistern filtration or "pre-filtration," as it is sometimes called, is an important part of any active water harvesting system's success. The less debris forced to pass through the hydraulic lift/distribution side of a cistern, the fewer pumps to fix, filters to clean, and valves to replace.

The more important reason to pre-filter has to do with a cistern's lentic, or still-water, state. It's a simple concept, really. If you prevent debris from floating in a cistern, fewer surfaces exist for algae, bacteria, and viruses to glom onto and grow, particularly during the many long stretches of time when the water in a cistern sits motionless.

Some forms of pre-cistern filtration work better in conjunction with pitched roofs, while others are used more frequently in flat-roof situations. In all cases, as the number and size of mature trees around your roof increases, the need for pre-cistern filtration grows accordingly. Tree branches, leaves, pine needles, bird poop, and other natural and man-made causes can pose problems if ignored during the conveyance process.

Even though significant pre-cistern filtration typically does not occur until later in the conveyance process, leaf screens installed along focal points can prevent debris from being diverted toward the cistern. Leaf screens attached to gutters and the openings of downspouts are an inexpensive and effective method of pre-filtration, but they rarely make additional filtration along the conveyance system unnecessary. Leaf screens go by a variety of names, including gutter guards, debris traps, and leaf catchers, among others.

Common leaf-screen materials include plastic, woven fabric, or wire mesh. They can be especially helpful in preventing birds, rodents, reptiles, amphibians, and insects from entering your conveyance system. Even though more debris collects on flat roofs than on pitched roofs, leaf screens are not recommended in conjunction with *canales* because these focal points can clog and back up quickly if inhibited by screening. This can cause roofs to leak and in extreme situations mold to form, which will take serious time and money to fix.

Smaller particles of debris can also be removed from your conveyance stream. These can be divided into two categories: "filters" and "first-flush devices." Although neither of these methods will prevent all particulate from entering your cistern, each technique can significantly increase the water quality produced by your system. If you have the inclination and the budget, you should consider applying a number of these techniques because redundancy is often very helpful when it comes to filtration.

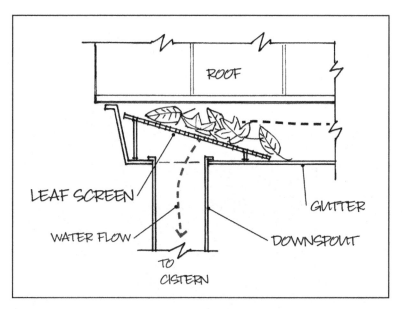

A Leaf Screen

Like leaf screens, filters for pre-cistern filtration have various names, such as inlet filters, sediment traps, and sand filters. Essentially, they are barrels that contain one or more filters or filtration materials. They typically connect to underground lateral piping, and some are self-cleaning.

Even while they clean your collected water, first-flush devices are not actually filters. They divert the dirtiest portion of a roof's runoff into a separate conduit or small tank.

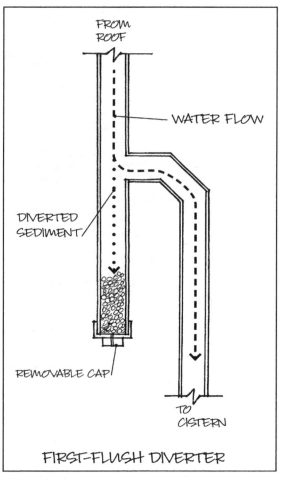

A First-Flush Diverter

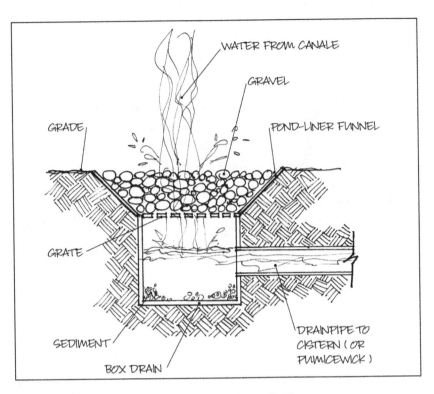

Box Drains Can Hold Back Significant Sediment

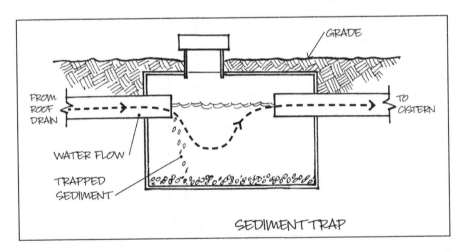

Sediment Traps Do a Better Job Than Box Drains

There are two common types of first-flush diverters. One device is the "dead-end pipe." Associated only with downspouts, not *canales* or lateral pipes, these diverters direct the first quantity of water (the dirtiest quantity) collected on a roof into a pipe that stops at a removable cap. The rest of the storm's runoff flows straight over the dirty, dead-ended water and into the remainder of the conveyance piping. The cap is removed after the storm and the dirty water can be drained into a nearby swale for use in your garden or landscape. Later the cap is screwed back on the dead-end pipe in order to be ready for the next storm event.

Another form of diverter uses a bucket, a counterweight, and a pulley to dump the dirtiest first flush out of the conveyance stream and into an appropriate place in the landscape. These "pulley pourers" are typically associated with collected water that has been focused in a *canale* or at the end of a gutter without a downspout. Like dead-end pipes, pulley pourers require the system to be reset after storm events in order to be prepared for the next storm, so these systems are not ideal for people who tend to travel during the rainy season.

Filtration between collection and storage can also be accomplished using a sediment trap or filter within the main lateral conveyance line. These need to be accessible so that regular monitoring and maintenance can occur. Cleaning out a sediment trap may not be the easiest way to chalk up gradual-greening time, but it can be one of the most important things you do when it comes to keeping the water in your tank clean.

Box drains can be used as sediment traps only for systems where small quantities of sediment are expected to accumulate or in systems where water quality is of little importance. Such drains are helpful, but a filter or trap designed specifically to hold back sediment is preferable.

Elevate the Discussion

From a cost perspective, cisterns are often divided into two species: aboveground and underground. Aboveground tanks are almost always less expensive than their subterranean cousins. The primary reason for this is the labor associated with digging a very large hole and a number of long trenches. But secondary reasons include systems design, materials, insurance, and equipment rental.

It is common for a 2,000-gallon underground tank to need a hole that is 10 feet wide by 10 feet long by 12 feet deep. This kind of excavation job almost always requires a backhoe.

Think of the effort involved in moving a small mountain of dirt one little scoop at a time.

The alternative of picks, shovels, wheelbarrows, hand tools, and sheer determination are not recommended. If it helps, just think of Tim Robbins and his "digging spoon" in *Shawshank Redemption*. If you're not familiar with the movie, just think of the effort involved in moving an earthen hill one ice-cream-counter-taster-spoon at a time. If you have hard, rocky, or heavy-clay soils, this procedure will not be for the fainthearted. Keep in mind that deep, wide holes are much harder to dig than shallow trenches because of the extra effort needed to lift soil a longer distance. (In other words, the removal of dirt from the depths of a vast hole to a centralized dirt pile represents more work than the removal of an equal volume of dirt to the side of a shallow trench because the distance from the trench to the side of the trench is relatively short.)

Further, the cost of underground conveyance piping is typically much higher than the shorter, unburied veins of an aboveground tank. In many cases, drop lines can be very short when the top of a tank is close to the roof. Add the cost of below-grade lateral lines that have to run out to an underground tank, and watch your installation costs rise.

The dirt removal associated with unseen cisterns is the most surprising cost for many people, especially in smaller-property situations where there is simply no room on-site for extra earth. Since much of the dirt removed from a 10 foot-by-10-foot-by-12-foot hole is not going back in due to the significant volume of the cistern itself, the remaining dirt must be dealt with in an appropriate way. Other than using the dirt somewhere on your property (by building swales, for example), the least expensive method is to call every building contractor in the phone book until you find one who needs fill dirt, and then tell the person that you have some that's free for the taking.

As far as predicting the amount of dirt that you will have left over from such a project, make sure you factor in something called the "swell factor." Whenever you dig a hole, the dirt you excavate increases in volume by about one-third due to the air that the excavated earth absorbs as it is removed from its naturally compacted soil. This can be an unforeseen problem for underground-cistern installation projects.

From an aesthetic point of view, the underground vs. aboveground debate usually translates into the difference between visible and invisible cisterns. The visible species stand on level ground or are partially buried in the ground. They can be screened by fencing or plant material or they can be enjoyed as proud, real-life symbols of sustainability. Invisible tanks are buried below grade, but the cistern's serviceway and air vent can often be seen if they are not screened or put out of sight.

The benefits of having an invisible cistern are considerable. First, if you happen to have a big, black industrial-looking, aboveground water-storage tank hanging out with you day in and day out, just off the front porch, from a curb-appeal perspective it can be harder to sell your house. Probably best to site it in the side yard or backyard and, if necessary, screen it with walls, fencing, vines, shrubs, trees, or some combination thereof.

If you're not willing to be regarded as an intriguing individual who happens to be ahead of his or her time with respect to the water issue, then sporting a front-yard cistern is not the way to go. Unless water issues in your area are already very serious, if you have the money or can get a fair loan, a tank of the underground variety is often worth the investment. (Of course, if you are willing to be so intriguing, I completely support you!)

A second benefit of invisible cisterns is that they provide more usable real estate. In close quarters, this is important, but on larger properties, it can be less of an issue. This benefit, however, is lessened in cases where privacy screening or a much-needed windbreak can be created with an appropriately placed aboveground tank.

Thirdly, in order to prevent freezing, below-grade cisterns are especially preferable compared to the above-grade species. Pumps, pipes, and the narrow-diameter fittings on the distribution side of the tank wall are particularly susceptible to frost damage. So, if maintenance costs are figured in, a leaky aboveground cistern can actually cost much more money and/or create more hassle than underground systems.

Even given these benefits, most cisterns worldwide are installed aboveground. Americans are an exception: being so rich (in the old days) and image-conscious (probably still), we tend to bury cisterns much more frequently. But in warm locales, like Tucson, Arizona, and Austin, Texas, visible cistern systems are gaining greater acceptance.

In Santa Fe, much of the populace happens to have a doubly green perspective. Here, where the frost depth is between 18 and 24 inches, cisterns are typically buried completely underground even though a significant cost savings can be attained by only partially burying a storage tank.

Aboveground cisterns, including partially buried tanks, can be used by anyone with a strong desire for a cistern and a reasonable budget. Most people, no matter what type of cistern they have, will need to figure in the purchasing of not only the necessary materials and plumbing tools, but also the hiring of a licensed electrician to hook up the pump, pressure tank (if desired or required), float switch(es), and/or pressure-sensitive switching system.

Another benefit of aboveground cistern systems is that it's not always necessary to use a pump to deliver water to plants. If your property has adequate slope, water can be drained out of the tank and diverted to plants, especially if these plants are appropriately placed. According to Santa Fe inventor Windy Dankoff, "Drip irrigation can function reasonably well on a PSI [pounds per square inch] much lower than the 10 to 15 PSI that irrigation-supply companies suggest." In other words, the head (pressure) from your aboveground cistern might make a pressure tank unnecessary.

Installation costs vary with on-site factors including slope, access, the number of *canales* or downspouts, existing vegetation, planned vegetation, and the number of float switches, recharge lines, pressure tanks, pumps, filters, valve boxes, and distribution lines. Connecting a cistern to drip irrigation can easily be done, but it's more complicated than most people think. The easiest way to pump from a cistern is to connect a sump pump to a hose. Unfortunately, drip irrigation is much more efficient than hoses and/or soaker hoses, but sump pumps don't work with drip irrigation unless the water first goes through a pressure tank.

In addition to determining whether you can afford an underground tank or not, in order to determine the best cistern for your property, it is helpful to have already developed significant portions of a landscape design. In the next section we will address some of the landscape-design issues associated with the cistern placement on your property.

Site Your Cistern Well

Before making a final decision as to the species of your cistern, it is often best to consider where to put the tank. In order to do this you'll want to consider the following three axioms of successful placement:

- Reduce costs by using gravity.
- Save money by understanding the physical characteristics of your tank.
- Increase your quality of life by making smart choices.

In order to reduce costs (and hassles) and use gravity to your advantage, be sure not to drop your conveyance lines too quickly. In the case of subterranean cisterns, lines that fall too quickly require digging an unnecessarily deep hole, and this costs time and money. In the case of ground-level tanks, you may have to settle for a shorter cistern than would have been possible had you dropped your gutter at the magic rate of 1/4 inch per linear foot.

In addition, significant cash can be saved if attention is paid to keeping the lengths of your conveyance conduit short. This is done primarily by the choice you make in determining the location of your cistern, but you should also consider the lengths of other conduits such as electrical lines, distribution pipes, overflow pipes, and venting conduits.

Some systems require one or more separate pump tanks in their conveyance systems. These small, temporary holding tanks detain runoff briefly so that water can be pumped as quickly as possible to the main cistern. In my view, the prospect of this additional level of complexity makes a good argument for incorporating some good old-fashioned passive water harvesting techniques like those described in chapter 1—instead of going through the trouble of installing a separate pump tank.

Sometimes, however, the physical characteristics of a piece of property will suggest and, more importantly, the financial budget of a property owner will allow for a distinct pump tank to be installed on a part of the property that is lower in elevation than the main cistern. If this is the case, it is important to be aware that this often generates a number of extra costs.

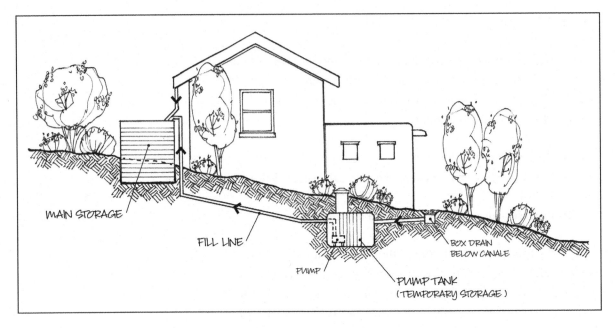

MAIN STORAGE

FILL LINE

PUMP

PUMP TANK
(TEMPORARY STORAGE)

BOX DRAIN
BELOW CANALE

Avoid Pump Tanks if Possible

You will need to buy more materials, spend more design time, pay for more labor hours, and consistently work to monitor and maintain the system. Extra materials would include another tank, a separate pump, wire and electrical parts, a float switch for turning the pump on and off, additional lengths of pipe, and associated fittings that will convey precipitation from the pump tank to the system's larger, long-term storage vessel.

Design time is critical and should be considered as an expense for this project. Unless significant dues have been paid in the trade, it'll take the typical gradual greener some significant time wrapping his or her head around such a system. The equation that must be solved is twofold. You'll have to determine what size pump will be needed to remove water from the pump tank at an optimal rate during typical storm events and also what size pump will be able to handle peak flows, that is, large runoff-water quantities that occur during short periods of time.

The largest up-front cost of using pump tanks instead of gravity will be the labor involved in installing all of these extras. Not only are electricians' rates (rightfully) considerable, but the labor associated with digging a trench in hard, compact earth all the way to the system's central cistern will also take a toll on your pocketbook.

The cost of which you should be most aware is the expense of monitoring and maintaining a whole new pump system. Understand that the rest of your system will probably be enough to keep most people pretty busy, so before choosing the pump-tank route, remember that low places on your property can be great candidates for a pumice wick, a Santa Fe drain, or perhaps a fish-scale swale system.

As far as your cistern size goes, at the risk of sounding like a pusher of the consumeristic mentality of "bigger is better," it makes sense to install as large a cistern as you can afford. Since the costs of collection, conveyance, filtration, lift, and distribution typically have little, if any, relationship to the size of a cistern, significant increases in a tank's dimension can have a surprisingly small effect on the overall cost of an entire active water harvesting project.

Determine the actual dimensions of your tank by asking your supplier. A 1,000-gallon cistern equals 133.33 cubic feet of volume. This translates into about the size of a light pickup truck. Typically, a 10,000-gallon tank will fit easily on a good-sized flatbed, and the largest single tank that can fit on an 18-wheeler is, according to my longtime water harvesting friend Richard Jennings of Earthwrights Designs, a 29,000-gallon cistern made out of a corrugated metal culvert. Jennings is quick to point out that he has stopped using culverts for cisterns until he finds a more reliable supplier for the bladders that go inside them.

The dimensions of your tank are important because you will need to know them in order to safely deliver and install the tank. Some materials require a crane, while others can be rolled around by one careful laborer. When large piles of dirt, a big cistern, a backhoe, hand tools, a dump truck for removing extra earth, and a crew of workers get added together, they take up a great deal of physical space, so be sure to understand how these various parts-as-a-whole will function efficiently during the installation process. If you happen to be near a water table or bedrock, you may need to look into low-profile tanks and/ or modular-tank systems so that you can collect the amount of water that you want to store without having to use dynamite or to forever worry that your tank is stable at the upper edge of a local water table.

Two other characteristics that can also be very important are the empty and full weights of your water-storage vessel. Will your cistern need to be installed with a crane or a backhoe, or can you slide it into place by yourself or with your crew? Will your ground-level cistern need to be tied down to prevent it, when empty, from blowing away and barreling down the street during a strong windstorm?

Will your full cistern settle inappropriately if it is not properly bedded underneath? At 8.3 pounds per gallon, in steep-slope situations make sure that a licensed engineer stamps any retaining wall associated with the installation of your cistern, and in soft-soil situations, make sure that an adequate foundation is prepared. A quick way to lose money after all of your efforts would be to underestimate the weight and potential buoyancy of a cistern.

The importance of careful planning includes making sure your installation project runs as smoothly as possible.

The importance of careful planning includes making sure your installation project runs as smoothly as possible. Essentially, this means you should do your best to predict all of the project's potential impacts. Underground tanks have serviceways and vents that protrude out from them from the top, so keep in mind that although an unsightly tank is hidden by the earth, these two items will remain potentially visible. Each can typically be hidden effectively with plant material, but they can look pretty ugly if not handled properly. Also, while keeping the tank inaccessible to children and anyone without the key or combination to the padlock associated with its lid, plan to make servicing your system as accessible and hassle free as possible. This will improve your quality of life. If the tank is accessible, you can monitor it more easily and identify any problems early on, and when you have maintenance to take care of, an accessible tank makes servicing the cistern a relative breeze.

Once you have determined the size, site location, and landscape design associated with your cistern, it always makes sense to run your ideas by other people, including the permitting agencies that may have jurisdiction over your project. Another set of eyes will often see problems and ask questions that you might have overlooked. For example, what will happen to the overflow water during heavy rains? What cost savings might you get from moving your tank closer to your electrical source?

These and many more questions will come up if you plan and design your system with the help of other people.

Professionals, of course, are preferred, but since so much of this is common sense, your friends or next-door neighbor might also become valuable sources of critical information.

Store the Rain

Cisterns come in a wide variety of materials, and each has benefits and drawbacks. Here is some of the general lowdown on a number of the most common materials. For all specific information, check with the supplier of the material and do whatever due diligence is required to ensure that the tank material you get is one that will work with your desires, your budget, and the innate requirements of your property.

Reinforced concrete tanks can be an economical choice, but with the rising costs of concrete, you will definitely want to get an accurate, up-to-date estimate for this and other materials. Concrete tanks can be prefabricated, typically as septic tanks, or they can be built on site to spec. Concrete cisterns have one advantage that is not true of all buried cisterns, which is that they usually can support significant overhead vehicular traffic. A disadvantage is that they are often coated with tar in order to prevent water from leaching slowly through the wall of the tank. Unfortunately, this means that tar will then leach into your water supply and end up in your soil. Although plants will survive with this kind of water, tar-coated tanks are not recommended for potable usages.

Ferrocement, or ferroconcrete, is a form of concrete tank that uses rebar and chicken wire. First, a tank-shaped carapace, or shell, is formed out of two ferrous materials and then a precise mixture of sand, cement, and water is spread, plaster style. Ferrocement is thinner than concrete, so this material can be the least expensive of all cistern materials. The key is that you have to be highly organized and efficient during the installation (i.e., on-site creation) of a ferrocement tank.

"Your mix has to be not too stiff and not too goopy," according to my friend Jeremiah Kidd, who owns San Isidro Permaculture, another Santa Fe–based permacultural landscaping firm. When we spoke over the phone, he was leaving for Uganda to help a clinic build a couple of ferrocement cisterns. "It's best to have enough people and enough materials to, basically, not stop until you're done. If you have to let the tank sit overnight without finishing a job, it's important to keep the next day's starting joint covered and consistently humid—but not too wet."

Ferrocement tanks can be installed as aboveground, partially buried, or underground cisterns. Art Ludwig's book *Water Storage* has some of the best material I have found on the subject, but see also www.ferrocement.com before renting that cement mixer.

Fiberglass cisterns usually need a crane during installation because large tanks are typically the only kinds of fiberglass cisterns installed. Fiberglass is usually not cost-effective until the tank is at least 6,000 gallons in size. Although lightweight, it's also a durable material that can be specified to withstand the mass of a loaded backhoe. Since fiberglass threads that migrate from the tank can have serious health impacts, especially on the digestive system, fiberglass cisterns should be thoroughly coated with a USDA-approved food-grade coating.

Metals are often avoided in tank construction due to their high price tag, but there are two exceptions to this. **Galvanized metal** and **corrugated steel** can be intelligent choices especially because metal can be cost-effective, particularly in larger sizes when specified to handle the pressure of vehicular traffic.

Plastic is a very common cistern material that can be used aboveground, below grade, or partially buried. It's lightweight so it's easy to deliver to a property, and it's easy to maneuver around a site. Plastic is a comparatively cheap material that can easily provide potable water.

Aboveground cisterns work best, as mentioned, when they're dark colored because this prevents algae growth within the tank. Sometimes a protective UV coating is required for this type of tank. If punctured, plastic can be patched a little bit more easily than some of the other materials, but since you want to avoid ever puncturing any water-storage vessel, this is not the greatest of its strengths. Due to the issue of collapsibility, large-capacity plastic cistern systems require that individual tanks be connected near their bottoms by a pipe or pipes. For many folks, this is a serious drawback. Who wants to worry about leaks occurring in the pipe connections between water tanks that are 10 feet underground? (A multitank system is what I chose for my home after talking with my wife, going over pros and cons, and so far, so good. Since my alternative was to crane a tank over my house and around a very large tree, I chose the relatively common multitank approach to water storage. The price was right and I was able to squeeze significant storage capacity into my backyard without getting too close to any of its old retaining walls. Plus, not only have my tanks and connections not leaked, I worry about them only infrequently.)

Another form of plastic cistern, known as a modular tank system, is also gaining momentum in the industry. Picture hundreds of super strong plastic milk crates stacked 10 feet deep under an entire parking lot. When these crates are wrapped and taped up like a gift, they make one of the most inexpensive and versatile water harvesting vessels imaginable, but the soft plastic liners used in these types of tanks leach more toxins than the harder plastic tanks that are widely available.

Plan an Exit Strategy

What happens during wet times when your cistern is full and more precipitation is being conveyed into the tank? The short answer is that it is released via an overflow pipe into an appropriately controlled part of your property. The long answer features a conundrum and fun-filled analogy.

One of the most important parts of a cistern system has no clear anatomical parallel to any part of a circulatory system. Like any mammal's bladder, a cistern must have an exit strategy for times when its tank is full. Since the endpoint of any urinary tract is its urethra, let's work with this analogy just for fun, shall we?

Off the top, a visible cistern typically has a visible overflow and urethra, and an underground tank has an invisible overflow pipe and urethra. In either case, it is important to point your pipe and urethra toward appropriate places. In the case of a cistern, it is usually essential to direct overflow water into a pumice wick, Santa Fe drain, French drain, swale, or controlled *arroyo* (e.g., with a properly specified check dam or gabion below) and some appropriate plant material. If this kind of attention is not paid, wherever your urethra is scheduled to discharge, it can set in motion soil erosion and property devaluation as opposed to aquifer recharge and localized soil building. Remember, an entire home's roof water will be focused at one particular place during an already wet time. This constitutes a great deal of force that many types of soil are incapable of absorbing efficiently.

The problem is what to do with excess water when there's already heavy storm precipitation and how to make appropriate use of the significant quantities of water that you expect to get when your cistern is full. Make the best of what nature gives you. For example, many riparian plant species thrive by being temporarily flooded. If overflow water can be stored passively in the soil, this kind of species often does quite well with what amounts to, in this analogy, your active water harvesting system's urinal.

You'll also need to find the right "daylight point" for your pipe. In the field of civil engineering, this is the point where an underground pipe pops out of a slope and discharges water continuously, seasonally, regularly, or randomly—without backing up or clogging. A big Santa Fe drain with rocks or rubble wrapped in a filter fabric can function as a kind of subsurface daylight point, but this is not ideal and can cause problems. On the other hand, if soil percolation is good and if your Santa Fe drain is sized large enough, overflow pipes to both true daylight points and subsurface ones will simultaneously function as a mini-aquifer-recharge facility during times when the tank is full and it's raining.

Provide Access and Ventilation

A serviceway, or manhole, provides access to the tank whenever it needs to be cleaned out, winterized, inspected, and/or requires servicing. Through a trap door leading directly into aboveground tanks, or via a short serviceway extension in the case of underground tanks, access must be wide enough for you or your average plumber to navigate.

In some ways the terms *invisible cistern* and *underground cistern* are misnomers because it is best to have quick, easy access to your serviceway aboveground, rather than burying it underground with the tank. However, if aesthetics are paramount (as they ended up being at our house for our cistern), you can bury your serviceway under mulch and make it invisible. Burying serviceways under soil, fill dirt, or materials that are even harder to get through (such as an asphalt driveway) is always a possibility, but access to your tank is important, so covering your serviceway in this manner is not recommended.

Some of the parts that you may need to service within

your tank are its pump, vent pipe, water lines, float switches, electrical lines, foot valves, check valves, and level readers. In many situations, you will need to winterize your system in the late fall or early winter, which also means you will need access in late winter or early spring. If your cistern system needs winterization within its tank, you will want to make regular access to the inside of the tank as convenient as possible; otherwise, you will inevitably neglect the simple chore and end up having to fix frozen pipes in the spring.

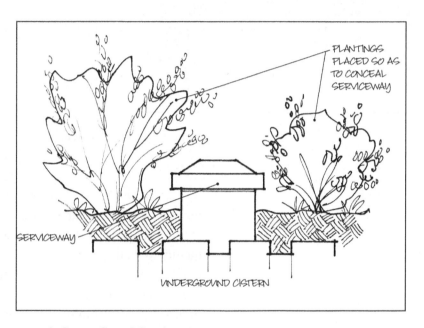

A Camouflaged Serviceway

Screening your serviceway can be accomplished with well-chosen, properly placed plant material, large boulders, benches, or other patio furniture. The most important thing to remember is to always keep your serviceway locked whenever it is not being used. People, especially children, *will* eventually fall in if they are not completely and permanently prevented.

Another important, but sometimes overlooked, aspect of active water harvesting is ventilation. Pumps run much better and cisterns function much more efficiently when air is allowed to flow freely in and out of the tank. Vents, upside-down J-shaped pipes, can be awkward to the eye. If possible, install your cistern so that your vent pipe can be placed out of sight.

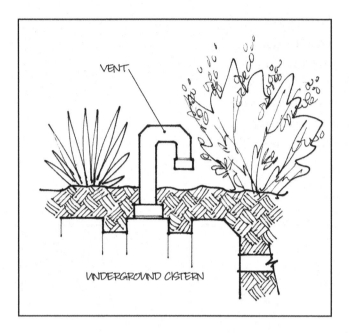

Your All-Important Vent

Be sure to prevent anything other than air from traveling through your vent pipe, from mice to mosquitoes. This can cause problems with everything from your cistern's water quality to your quality of life. If water is ever observed flowing out of your vent pipe, your overflow pipe is probably clogged, and if water is traveling into your cistern via your vent pipe, in order to prevent runoff soil from entering your tank, you will probably need to raise your vent pipe before the next significant storm event.

Admit Your Inner Nature

In the spring of 1988, six months after moving to Santa Fe, I took a job at the Plaza Bakery located at the southwest corner of the centuries-old plaza. One slow day I started calculating the number of ice-cream items you could order at the counter: 20 flavors, 17 potential toppings, three kinds of cones, five sizes of cups—the number was astounding even before you considered double and triple scoops or all of the beverages and baked goods that you could work into an order.

Similarly, options abound with regard to cistern systems. We've discussed many different flavors of roof in both the pitched and flat varieties, three or four kinds of conveyance piping, a variety of pre-cistern filters, two or three cistern species, six or seven cistern materials—the number of possibilities boggles the mind even before you start talking about all of your optional bells, whistles, and, if you will, *toppings* that best fit your taste, lifestyle, and family budget.

The most noteworthy option confronting all cistern-system designers is whether or not to install an automatic auxiliary fill line for use during dry times. The most common alternative to the automatic fill line is the manual garden hose. When times are extremely dry, neither of these may be an option, but as we transition from a conventional view of our water supply to a more precipitation-based view, we should always try to maintain a backup water supply for our necessities during dry times, prolonged drought, electricity outages, municipal water shutoffs, or natural events (that can turn the tap off, such as blizzards, earthquakes, hurricanes, and landslides). This is the redundancy principle at work in water harvesting: Try not to design systems that are dependent on only one source of water.

If you choose to have an automatic fill pipe (and your municipal water system is functioning), then you're good to go. Your system will require a float switch that communicates to a valve that delivers an irrigation cycle's amount of water into the tank when the cistern is empty or very low. An alternative to this is a pressure-sensing switching system that can add more water whenever low water pressure is evident in the tank. Both serve the same function, but each can get accidentally stuck in the "On" position and waste a tremendous amount of water. For this reason the cherry on top of a cistern system with an automatic fill system is a bright red alarm light that flashes as a siren blares in order to let you know that your fill line has been left "On" for too long.

If you choose, as I do, to fill your cistern with a hose, and if you happen to be easily distracted, as I am, try to build a backup plan into your system that can prevent the wasteful amount of water that can be lost if a hose is left on and forgotten about. In our case, our backup water comes from a slow-producing well that makes it almost impossible for us to add more than 500 gallons of water to our cistern system in any given 24-hour period. Given the amount of vegetables that we grow, however, we sometimes have to rely on another layer of redundancy, city water.

I'm not sure if I'd want to admit the number of times I've forgotten to turn off a hose, but fortunately I can't remember. It happens. What you want to make sure of is that it does not happen for very long. Fortunately, I've found that my brain must have some file-checking program that kicks in during REM because I've never slept through the night with the garden hose on. Something always bolts me out of bed, less than half naked, into the dark of the backyard to crank down the valve before it becomes ridiculously too late.

Personal tendencies toward bouts of midnight paranoia aside, I wouldn't recommend relying on such a subconscious system of file checking. Instead, I suggest you choose any one of the following techniques:

- Put a rubber band around a wrist, a toe, or any extremity of your choice.
- Stick a penny or screw in your sock, shoe, or hat.
- Mark the back of your hand with "WATER," "W," or just plain old "!"
- Tape a big note on your bathroom mirror or the fridge that says, "Turn it off, NOW!"

For those of you who prefer not to have people ask you why you've scrawled on your own skin, you may just want to avoid pens and keep rubber bands and/or small, sharp wood-screws handy.

Like some of the choices at the bakery, the form of auxiliary-fill technique that you choose for your cistern system tells a fair amount about you as a person. The person who orders one scoop of sorbet in a cup is on a very different trip than the person who orders a three-scoop, extra-fudge sundae with everything. The person with the rubber band around his finger has a very different relationship to his cistern than the person who depends on an automatic fill line.

Automation is a wonderful thing if you can afford the costs of installation and the extra expenses associated with maintaining the technology. So it goes with supplemental water. Systems design, planning, purchasing, plumbing, and the electrical work typically needed for automated systems not only means more expense up front, but it also means more parts requiring cleaning, tweaking, fixing, or replacing.

The cost of water can also be greater in automated situations because people might not realize that their system has malfunctioned until they get a water bill and suddenly realize a valve from the local utility must have been stuck on for a month or more. At the other extreme, people living on well water could theoretically (if they never measure the water in the tank or check out the cistern's overflow point) overlook that their cisterns are being filled more by a local aquifer than by rain. The karmic costs of this kind of oversight would be of bank-bailout proportions. Still, I tend toward my manual hose technique, feel comforted that mine is a slow-producing well, and fervently pray that we never leave for a long weekend with the city hose left on.

When it comes to reading the level of water in your cistern, there's another option that tells us a great deal about how purely austere or lavishly automated you are. On the low-tech side, you can use a long dipstick to figure out how much water you have in your tank at any given time. On the high-tech side, you can buy a digital level reader. Mount it near your distribution system controls, and you're done. However, this minor step does add yet another line item on the cost of these systems, so keep in mind that a wooden stick with indelible ink markings on it will fit within any budget even if digital water-level readers do not.

Another choice often must be made: whether you connect your water line to a pressure tank or just let the pump run straight out of a hose whenever you turn it on. For drip irrigation, pressure tanks are usually needed except in special situations. When using a garden hose or an inexpensive (but less efficient) hose sprinkler system, no pressure tank is needed.

Last but not least, you must consider where to house your pump. The pump kingdom is divided into two species: the submersible pump and the inline pump. The former lives inside the tank itself and can reduce your installation expenses. The latter lives in a separate pump house and can cut maintenance costs. Where to put your pump? First you have to decide what kind of pump is right for you—vanilla or "Cherry Garcia" on a waffle cone topped with whipped cream, sprinkles, nuts, and hot fudge?

Automation is a wonderful thing if you can afford the costs of installation and the extra expenses associated with maintaining the technology.

Pump It Up

In the best of all possible worlds, you could simply turn a valve to "On" and watch gravity-fed distribution lines irrigate your plants. Unfortunately, existing site conditions, enthused landowners, and helpful cistern-system designers rarely end up on the same patch of dirt at the same time, so most people use an electric pump, in lieu of gravity, to move stored water.

If you happen to exercise more than most people, manual pumping is a good idea, especially if money is tight and water needs are low. Physically lifting significant quantities of water translates into a real commitment to regular exercise. Even though the hydraulics involved are much easier in a nearby cistern compared to a deep underground well, gradual-greening time spent in a backyard "gym" (designed more for the purpose of pumping water than iron) would be frequent and extensive. Still, the dream I have for my home (OK, and my quads, too) is to connect an exercise bike to our underground cistern, cover it with a small greenhouse, and then connect the bike to a pump that lifts water to a cistern for irrigating both plants in the greenhouse and those in the existing food forest that wraps around our property. I even fantasize about the construction of weight-lifting machines that bucket water instead of heavy useless dumbbells up and down. Unfortunately, these kinds of systems are not one-size-fits-all: they are difficult to design and produce inexpensively, and they're certainly not the kind of system you can find on eBay or Craig's List. The best design that I have seen comes out of Austin, Texas, through the work of an engineer-cyclist named Larry Glig, who can bike right up to a pump connection in his yard, and as he begins to pedal, he begins to spray water from his cistern and onto his landscape.

Solar-powered pumps are cost-effective if—*and only if*—the power grid is too far away to justify the cost of extending an electric line. During this era of relatively cheap energy, in most parts of the developed world grid-tied electricity is nearby, and a well-designed pump system will cost little to operate compared to the carbon footprint left behind by such machines.

Fortunately, my friend and fellow Santa Fean Windy Dankoff invented a famously effective line of solar pumps that do the job of lifting water with the sun's energy. "If you want a pressure-on-demand system, a battery system is required. These systems cost $1,000 and up," he said. "They are economical if it will cost that much or more to extend conventional power to the pump."

If you need only very low pressure for distribution, Dankoff says, some of the least expensive solar pumps will work. What's even better is that they are only a few hundred dollars more expensive than a conventional pump, so the utility-bill payoff is quicker and their carbon-foot print is significantly daintier than the non-renewable-fuels motif.

In power-grid situations, many drip irrigators' favorite pump is the Grundfos MQ. It's got a pressure tank built right into it, so it eliminates the need for connecting a separate pressure tank. The motor varies in speed to avoid rapid on/off cycling that would otherwise occur with a small pressure tank. This saves installation time and reduces the system's need for space, but on a materials basis, this pump costs more money. MQs cost a couple of hundred dollars more than the mass-produced single-horsepower jet pumps found at your local hardware store plus a typical six-gallon pressure tank. If the extra space for the pressure tank and a few basic plumbing skills are readily available, it may save money to use a standard jet pump instead of an MQ.

Whether you buy an MQ or you choose a more common model with a pressure tank connected further down the line, your pump will need to be housed somewhere. Underground pump houses, aboveground pump houses, and pumps located in utility closets, garages, basements, and toolsheds are all excellent options depending on your circumstances. Just try to have a way for water to drain easily out of your pump house in the likely event of someday having a leaky pipe. Such leaks will destroy your pump and many of its associated components if a flood should inundate your pump house.

If all this business about pumps is confusing, "do not hesitate to call a professional pump installer," Windy continued as his eyes tightened. "And let me warn your readers that 99 percent of plumbers know nothing about installing pumps, but 98 percent of them think they do. Conventional plumbers deal with water that is already pressurized, but the intake line of a pump follows a whole different set of rules that they rarely consider or even know to look for in the instruction manual."

There are four forms of distribution: hose watering, drip irrigation, bucket dumping, and the famous, but misunderstood, rain barrel.

Shallow-well jet pumps (nonsubmersible pumps also called "surface" or "inline" pumps) require more accoutrements than submersible sump-pump systems. The most complicated shallow-well systems contain over two-dozen parts, while submersible pump systems feasibly have just three: a hose, a pump, and a working electric outlet.

To take on the installation of any pressurizing system, most people will want to hire a professional. If you happen to be like most people, make sure that your cistern-systems contractor understands your level of desire for automation—that is, you probably want plenty of it. One of the best reasons to install an inline pump is so the pump can be easily accessed when it breaks down or needs maintenance. Should you choose to have someone else install a pump, "you should buy the pump from the installer," Windy's words breezed by with authority, "so warranty issues and contractual responsibilities are clear."

The main drawback of shallow-well pumps is the added costs associated with building a pump house, which is often located underground. If you plan to distribute your water with a hose, you should consider a sump pump because then you will not need a pressure tank to run your hoses. A bare-bones submersible pump system can consist of the pump, an electric outlet, a hose, and spray nozzle, so not only is your expense side of the ledger lowered by the sump pump option, your list of items to troubleshoot in the event of a problem is drastically reduced. What you'll often lose is the convenience of automation, so just keep in mind the concepts discussed in the previous chapter about the dialectic between manual systems and automated systems.

Roll Out the Barrel

As far as cistern-oriented landscape-distribution systems go, there are four forms: hose watering, drip irrigation, bucket dumping, and the famous, but misunderstood, rain barrel. You see, rain barrels are not really cisterns. In some of the same ways that toddlers aren't quite adults, rain barrels are a lot smaller than cisterns. They get knocked over easily, get quickly affected by the elements, and can tumble down hillsides faster than Jack and Jill. Most of all, they spill much too much.

Due to the excessive quantity of their spillage, rain barrels need to be considered first as a distribution technique rather than a storage component. Here's why: Since a 500-square-foot roof can yield 3,250 gallons of water in a year of average rainfall in a high-desert region like Santa Fe—enough to fill a 50-gallon drum 70-something times—few people with a great deal of free time are clairvoyant enough to get out there at just the right time to get the water out of said barrel.

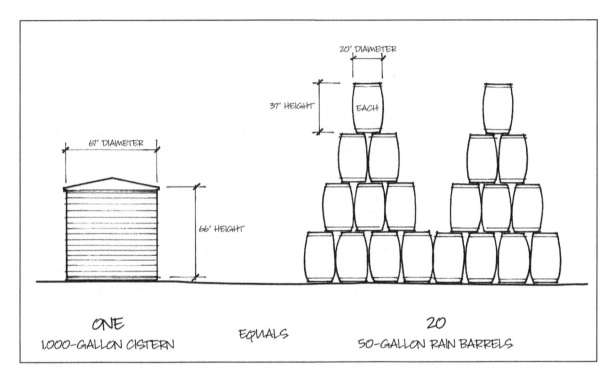

Cisterns vs. Barrels

Unless your rain barrel is well situated above garden beds that you would like to water, you will want to purchase an electric pump. After all, a full 50-gallon container weighs 415 pounds, so without a pump, distribution can devolve into a dangerous and destructive bout of pushing, shoving, inexact pouring, and relatively uncontrolled dumping. A five-gallon bucket, on the other hand is manageable, but this would mean you would need to make about 700 trips around your yard during the course of a year—most of which would take place during extreme downpours!

Other drawbacks to rain barrels include their tendency to attract mosquitoes and their often unaesthetic visual impact on a landscape. Unless you spend effort making your barrels look pretty, they will look like the glorified, extra-large garbage cans that they are. Some folks say a Santa Fe woman, in a valiant effort to save her struggling cat, actually got stuck and drowned in a rain barrel. Others suspect the husband of foul play and the most evil form of wife dumping one could imagine, but I digress. . . .

Whenever money is tight, your site is suitable, and your lifestyle is conducive to rain barrels, the simplicity of a sump pump dropped in a barrel and attached to a garden hose is one of the most beautiful sights in the world. So, at the beginning of your course in gradual greening, as a way of getting used to harvesting the rain, feel free to make use of rain barrels. They are inexpensive, educational, fun to use, and certainly much better than nothing—especially if you don't mind getting wet.

One of my most recent clients, Bette Booth, is the proud owner of five rain barrels, and it's unclear if she'll ever get rid of them. Her lot is small, but her desire to harvest precipitation is fierce. We tried to figure out a good place to stick a cistern on her property, but the only places that were big enough were either too close to the foundation of her house or too close to all of the underground utility lines coming into her garage.

"The Bioneers Conference of 2008 was transformational for me," Bette said over locally grown hot tea. "I returned convinced that I wanted to live in a full circle, and I realized that my water rights—rights that people were fighting for around the globe—were over my head."

At some point even Bette might want to roll her barrels down to her neighbor's house and teach him or her how to use them. By that time, perhaps the work we did to hold water in her soil will be doing a sufficient job so that she can still feel deeply connected to her local water cycle, or maybe she'll just bite the bullet and install the cistern she's always wanted. Like toddlers' clothes that get passed down through friends and relatives in any community, rain barrels can last a really long time and could be given (or sold at a garage sale) to those whose gradual-greening regimes are younger and less fully developed than your own.

Distribute Your Harvest

Not until water is intentionally distributed to some useful purpose is the process of water harvesting complete and its promise fulfilled. All of your efforts to collect, convey, filter, store, and pump precipitation will be for naught unless your water ends up at a desired place.

Rainwater can be put to any use inside or outside the home, but as I explained at the outset of chapter 3, my belief is that our culture will be much more willing to accept the idea of using rainwater inside of our buildings after having a decade's worth of experience with the majority of our cistern water being used for irrigation and other outdoor purposes.

Among the several forms of landscape irrigation, most of my experience has been with flood irrigation and drip irrigation as opposed to pop-up-sprinkler irrigation, central-pivot irrigation, lateral-move irrigation, and others. For anyone who has the time and commitment, I'm a big fan of hose watering, but my all-time favorite form of irrigation is one that one of my clients developed for the cistern we installed for him. I call it "the big-gulp cup system," but we'll get to Richard Word's trough-and-bucket-dumping system in a minute. Now, let's take a look at the best kind of irrigation for most residential and small-scale commercial applications.

Made up of small-diameter black tubing and hard-plastic water emitters, "drip," or "trickle," irrigation is a water-wise, time-efficient, and cost-effective way of irrigating plant material. Also known as "drip irritation," it consists of numerous parts from tees and couplings to valves and emitters. Compared to most systems that are used in the modern world, it should be noted that drip parts, tubes, and connections break far more often than their counterparts do in, say, outdoor lighting. What's worse is that the leaks associated with these frequently breaking parts can go unnoticed for relatively long periods of time, so systems should be regularly monitored and maintained. On the bright side, drip-irrigation parts are relatively inexpensive, and working with them can always count toward your gradual-greening time.

Professional drip-irrigation systems require two very different kinds of work. The infrastructure part of the project typically starts with a licensed and permitted cut into a main waterline. From this incision, a tee diverts water to a backflow-prevention device, pressure regulator, filter, and any number of automatic valves connected to a computerized, low-voltage timer. Backflow-prevention devices are typically not necessary with cistern systems as long as an air gap exists between the cistern and incoming supplemental water. At the appropriate time (often midnight in the summer to prevent evaporative loss and midday in the winter to reduce the risk of lines freezing) one of the valves opens and allows water into the surface "drips," that is, distribution tubing and drip emitters.

Surface-drip tubing comes in 1/4-inch increments up to 1 inch in diameter. In most residential situations, a combination of 1/2-inch and 1/4-inch tubing works perfectly. Not the prettiest of landscape features, drip tube is best buried by mulch or a few inches of soil. Emitters are precisely placed such that water dribbles onto the plant's upper-root system, making optimal use of the water distributed—especially if used in conjunction with Reese Baker's rock tubes that we explored in chapter 2.

Although emitters come in various sizes, I try to stick with one-gallon-per-hour emitters because it's easier to determine your system's expected water usage if all that you have to do is count how many emitters are associated with each of your zones. If your budget allows, install two emitters per plant. This will keep your plant alive if one of the emitters should clog, and it will help your plants grow symmetrically as long as you put the two emitters on opposite sides of each plant.

Fixing a drip irrigation system's surface drip lines and emitters is much easier and much less expensive than fixing broken parts of the system's infrastructure. Every able-bodied adult has the potential to learn and do the former, but not everyone has much of a chance of acquiring the plumbing skills necessary for the latter. Likewise, in climates that freeze in the winter, seasonal turn-ons and turn-offs can pretty easily be accomplished by any adult on a tight budget. Otherwise there are local irrigation and/or landscape-maintenance companies that will service your system in the spring and fall so that everything works as efficiently and productively as possible during the growing season.

Windy Dankoff, the eco-tech inventor, hunted me down during the final fact-check phase of writing this book, and I'm really glad he did. Windy has experimented with using gravity to water plants using standard drip irrigation parts in not-so-standard ways. As a result, he's got lots of wisdom to share.

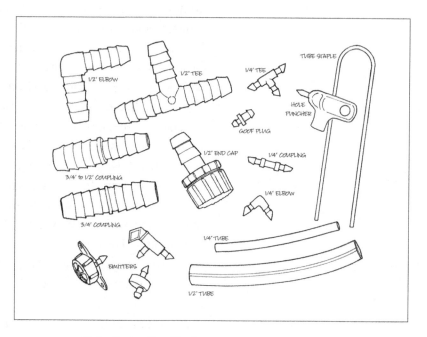

Drip Irrigation Parts

"I'll tell you how to get good performance from very low pressure, even just a few feet of drop," Windy told me while standing in the shade of an old but very functional solar panel. "First, avoid air pockets in your line, or the pressure of the water above might not be enough to push the air out of your half-inch tube. You'll get an air pocket stuck in there that will block the flow and drive you crazy. One of the best ways to do this is to make sure your line is always going slightly downhill. Avoid sending your drip tube back up the hill if you are trying to water using gravity, particularly if your slope is gradual and your lines are long.

"You want to use emitters that are rated at a greater flow rate than you actually want unless your tank is 30 or 40 feet in the air. In order to get more even distribution in the lowest parts of your system and at the beginning of your line, you can also use slower emitters in these places and get more uniform coverage."

The most important information that I got from the interview is that using drip irrigation at below-standard pressure may require some creative tinkering. "Ultimately, you can run drip irrigation with as little as 2 feet of elevation," Windy went on, "but you'll need to do lots of experimenting with the emitters and sometimes not fill your holes with emitters at all. T-Tape® is very handy for garden rows. It has a second passage inside that equalizes the pressure from one end to the other. I've seen it working on 2 feet of drop, as a rain barrel emptied."

Sediment may accumulate in low-pressure systems because of the low flow rates. Windy said "You will want to filter your water as much as possible, and it may be good to install a hose connection so you can flush the system with standard pressure from time to time."

Windy also warned that you cannot use ordinary automatic and electric valves that work with standard drip irrigation. "They run for a year on a little battery because they use water pressure to help power the valve open. Without 'normal' pressure, they don't work. You'll need to use a motorized valve for low pressure or to find other strategies to semi-automate, like manually filling a small tank and just letting a prescribed amount of water run out," he said.

We mostly use drip irrigation on our property, but especially after a wet winter, I sow peas in the vegetable garden, and these require hose spraying. I like to slip out as the rosy-fingered dawn slips over the southeastern corner of the Rocky Mountains, and for a brief moment I listen to and acknowledge the syncopated chorus of birds chattering like mad in the neighborhood. While much of the world wakes up, the rest of it, the nocturnal part, prays for some drowsy prey and a quick snack before bed. Hose watering also turns out to be an auspicious way to get some gradual-greening time in to any work day, that is, first—before those millions of distractions that we get each day start falling all around us.

Most people think they are too busy to water plants with a hose, and many folks truly are. But it's easy to imagine during times of high gas prices, high food prices, and high unemployment that the average daily time spent commuting to the gym would be at least between eight and 10 minutes for most Americans. If you take your time, your exercise, your garden, and your precipitation seriously enough, you can get a reasonable workout running around watering a small garden with a heavy hose. Add the planting, mulching, composting, and swaling that I suggest above, and we're talking about building up a pretty nice aerobic, whole-body workout over the course of a gradual-greening experience.

Unlike flood irrigation and spray irrigation, given a thick mulch, drip irrigation and hose watering lose virtually none of their moisture to evaporation. The flood-irrigation technique used in the *acequia* systems of northern New Mexico is often criticized for its waste, but because rivers naturally flood their banks, it is certainly healthy for a river to flood in this way especially if we are, as a culture, going to try to control rivers and prevent natural flooding in the way that we almost always do. Unless your cisterns, your rains, and your collection surfaces are all ample, flood irrigating is not recommended from cistern systems because precise water placement is usually a critical part of any cistern system's water budget.

The closest thing to "flood" irrigation that you might want to apply with a limited supply of water is the big-gulp cup system that I mentioned above. It was developed by a former client and lawyer-friend named Rick Word. He and his green-thumbed wife, Laura Brown, take turns opening a valve about 6 feet below the bottom of their aboveground tank. Subsequently, water fills a small metal trough. From there, they take large cups (which might be said to be the size of small buckets) and scoop water out of the trough. Then, on an almost-daily basis they simply walk over and pour a little life into their happy, healthy garden. This is pure water harvesting elegance at its best, but only because the right combination of site, client, and designer/contractor happened to appear on the same patch of dirt at the same time.

Some vegetable gardens get all the luck.

Add the planting, mulching, composting, and swaling that I suggest above, and we're talking about building up a pretty nice aerobic, whole-body workout.

4

Wastewater Harvesting Methods

Talk about your plenty.
Talk about your ills.
One man gathers
What another man spills.
—Robert Hunter, "St. Stephen"

Recycle the Rain

Whether the water you consume comes from a cistern, aquifer, surface-water supply, or desalination plant, it can be reused on your property again and again. First, harvested precipitation can be treated with a bank of micron filters and a pair of ultraviolet lights (see next section) such that the water can be safely consumed in drinking faucets and via shower heads. Some of the collected precipitation could be diverted to dishwashers, washing machines, and flush toilets.

After becoming one of two types of water—greywater (waste from bathroom sinks, showers, and clothes washers) or blackwater (from kitchen sinks and toilets)—these resources can be treated in any number of ways. Most people would prefer it go outside to enhance the landscape, but it is certainly possible, depending on your level of scrupulousness, to filter wastewater to the point of being reused in any and all of the fixtures in the house. On-site water treatment has the potential to effectively close a home or a community's water loop to a point at which harvested rainwater would be used only for drinking, food crops, and other forms of irrigation. Every other residential and many commercial and industrial uses could be taken care of with water cycling through the system indefinitely in a way that requires only rare replenishment.

One of the best examples of architecture that supports this level of water consciousness is the Earthship technology that Michael Reynolds has developed over the course of many decades. Built primarily in and around Taos, New Mexico, Earthship technology has been applied in nearly every climate on the planet. Reynolds's homes and subdivisions, which are constructed principally of old tires, used cans and bottles, earth, and concrete, use roof water for drinking, washing, indoor gardening, and toilet flushing before what's left heads out to the landscape. "I used to be an architect," Reynolds says as he hustles determinedly across a video screen at www.earthship. com, "now, I'm a biotech." Suddenly, the sound of a loud motorcycle and an image of the long-, wavy-, and white-haired Reynolds speeds past. Then, in the distance, you catch a longer shot of the eco-home builder gunning it down a lonely winding road in the desert.

"Architecture is not responding fast enough to the needs of our planet and her people," Reynolds said in a 2009 phone interview. "In fact, 21st-century architecture is still mostly reducing our chances of survival as a species. Biotecture addresses the reality head on. People who live in Earthships have no utility bills to speak of, and the buildings themselves are made almost entirely out of recycled materials and dirt harvested from the building site."

The water systems that Reynolds's homes use are particularly efficient. In his book *Water from the Sky,* Reynolds describes how roof water is collected in cisterns and is directed first to the sinks, showers, and washing machines inside the home. After being used, the resulting greywater drops into a planter bed that filters the water such that it can be used for a third time to flush the toilets in his homes. Finally, this water is used in planting beds for a fourth time just outside the Earthship.

"Although Earthships have at times fallen short of some of their goals," admits architect and former Earthship builder Peter Wilson, "the bottom line is that they have been designing, experimenting, and working to achieve the goals that mainstream architecture is only now coming to grips with. Given the obvious threats of global warming, fortunately Michael is no longer one of the lone voices in an 'alternative' green building industry. Energy-neutral, passive-solar structures that harvest and reuse precipitation must become the norm if we are going to prevent climate change in any meaningful way," Wilson, a Santa Fe–based architect at Net Zero Design, continued as he pointed to his newborn baby. "His future depends on it."

As a society, we are nowhere near the point of accepting Earthships into mainstream culture, but someday, perhaps only after a few more evolutions in the home-building industry, we will. My guess is that only after 10 good, solid years of harvesting cistern water for landscaping would we expect any large housing market to fully accept the use of rainwater inside the house. Only after another decade, could we expect to see building codes that allow the reuse of blackwater in a closed or near-closed water loop. The amount of time for this evolution to occur will always depend on the seriousness of the local water (and/or food) situation, any given community's desire to develop and grow, and the charisma and success of people like Reynolds and Wilson. For most people, however, this is an evolution that will be many years in coming.

Although it is theoretically possible for a community to become aquifer independent on roof water alone, water sustainability is not likely to occur without the appropriate reuse of our wastewater. Since sewage is so readily available and since the human need for water is continuing to increase, it is hard to imagine a future without much more effluent cleansing and on-site or downstream reuse. Call it what you will—water purification, treatment, filtration, sterilization, or recycling—along with being the universal solvent, water is particularly cleanable and, therefore, worthy of being collected, distributed, and reharvested indefinitely.

Since every community has its own unique, local characteristics it is clear that the pending water revolution needs people not only with a little expert knowledge, but also with a reasonable understanding of how things get done in particular communities. The world of wastewater harvesting is a dicey one, filled with slow bureaucrats, stigmas associated with "waste," and a serious responsibility to guarantee public health, but the reward of obtaining clean and fresh water is well worth the risks involved. The bottom line is that the wastewater harvesting industry resembles a rising tide that both tree-hugging eco-maniacs and forward-thinking entrepreneurs and investors would be wise to wade into.

The pending water revolution needs people not only with a little expert knowledge, but also with a reasonable understanding of how things get done in particular communities.

Kill Anything Serious

Due to the decades it will take for most of mainstream culture to accept a near-closed water loop, just as we focused in chapter 3 on water harvesting for plants in the landscape, we will have a similar focus here. People are simply not ready to drink sewage—or even have it filtered and pumped back into the toilet. It's not only that I don't want to lose readers by promoting a societal taboo, it's also personal. The idea of drinking effluent-cleansed water disgusts me about as much as it disgusts you.

Even though nobody you know ever poops on your roof, birds do. Squirrels do. Even mice and worms do. So, at least for many government regulators, there is a real worry that people could get sick from any E. coli that gets washed into a cistern from a roof. In other words, it's very unlikely but certainly possible that your cistern water could turn dangerously septic and in fact have an equivalent effect on your health as drinking sewage itself.

For this reason, before cistern water is distributed inside the home, especially if the water is to be used for drinking purposes, it would be wise to consider the possibility of a worst-case scenario occurring in your tank. Just visualize a family of rodents crawling into your cistern and drowning. Or maybe a flock of migrating birds all using the facilities on your roof before heading south.

In these scenarios, according to Stephen Wiman, president of Good Water Company in Santa Fe, "after ozonating your water in your cistern, hit it with a bank of three filters—a 20-micron filter, a 10-micron, and then a five," Wiman explained one afternoon over an order of curly fries. From the patio at Counter Culture, we could see his office door on Baca Street, but his new building obscured any view of his brand-new 20-foot-tall cistern. "The purpose of this is to eliminate particulate that might create a shadow effect in the UV."

The ultraviolet-light component is a two-light bank of fragile UV tubes, 28 inches long. As water floats past the long bulb of pseudo sunshine, Wiman continued, "it gets blasted by a light that doesn't make fecal organisms disappear—it neutralizes their DNA and, basically, it'll kill anything serious." If pieces of matter get in the way of dangerous particles in the water, unhealthy bacteria and viruses will not be duly scorched, so the bank of micron filters are as important as the lights themselves.

At the same time, Wiman stressed that regular testing would be wise and the option of treating roof water with some of the same chemicals that conventional water utilities use is always available, though he admitted, "My regular water-treatment clients usually want me to get those substances out of their drinking water." Another way to keep your redistributed precipitation clean, he said, is to avoid the use of carbon filters because "carbon is a haven for bacteria, and it's mostly used to filter out chlorine, which isn't found in rainwater."

Wiman's initial reference to "ozonating" refers to the installation of a floating bubbler in your cistern. "This adds air to your stored water and keeps it from stagnating," he explained.

Ultimately, if times ever get really tough, the only thing you need to know now is that sewage can be treated to this level of cleanliness in pretty much the same way. It's not something I recommend because of the ample quantities of clean rainwater that exist for the drinking-water needs of the human race, but it is certainly an option. As folks all over the planet know by experience, on the one hand many people are involuntarily drinking treated effluent, just as space-station astronauts spend days drinking their own treated pee with space-shot glee.

It may be nauseating to think of drinking the same water that is flushed down your neighbor's toilet or that which has been marinating with rodent bile, but whenever treatment is adequate people do not get sick from liquid wastes. After ensuring that any given treatment system works, the key is to constantly monitor the system and to fix any problems immediately as they occur. Still, we are a long way off as a society before backyard blackwater systems that send water back into the house come into vogue, so the remainder of this chapter will be focused on ways to improve your landscape *today* while harvesting wastewater.

Reuse Greywater

As water resources continue to dry up around the globe, in many cases municipalities, water authorities, utility companies, and even residential well users will have no choice but to reduce and/or eliminate the use of conventional water sources for landscaping. In addition to using active and passive water harvesting techniques, an excellent alternative is greywater for watering plants, trees, and compost piles.

In the broadest sense greywater includes all nonindustrial wastewater except for toilet water. However, due to the difficulty of filtering disease-laden fats, oils, and grease, kitchen-sink waste is usually dumped into the same category as toilet water, both of which are considered blackwater, sewage, sludge, crap, and the 10,000 names for untreated effluent.

As long as diapers, liquid fabric softeners, and pollutants such as gasoline are not included in the wash, laundry water can be the best source of greywater to access, but wastewater from showers and tubs is also an acceptable source. Bathroom sinks are generally viewed as the safest sources of greywater.

The definitive reference materials for greywater have been written by greywater guru Art Ludwig. Available at www. oasisdesign.net, Ludwig's books incorporate many years of experience in the field. Although greywater recycling is not high-tech science, significant mistakes can be avoided by doing a little research in advance by simply reading his clear, detailed, and comprehensive condensation of the tests, tribulations, and triumphs of a vast number of greywater systems. (He also has written a very informative book called *Water Storage*, which I highly recommend.)

My friend and colleague has been especially giddy about his new "laundry to landscape" system, which Ludwig says is "the cheapest, simplest, most effective way to distribute laundry water to plants." As such, it appears to be a great candidate for a weekend gradual-greening project. Specific plans can be found at www.oasisdesign.net.

Contrary to what most people think, greywater should not be stored in a tank because it turns septic quickly. This is something to avoid not only for health reasons, but also because the tank will start to stink and become a rancid, hard-to-clean and difficult-to-remove liability on your property.

Unless cleaned by an expensive sand filter, greywater is also not a good candidate for use in drip irrigation. Because of the particulate in greywater, emitters will get clogged, filters will need constant cleaning, money will be wasted, and, according to Ludwig, the typical relatively inexpensive drip-irrigation systems associated with greywater are typically abandoned in five years or less.

In my experience, the best way to divert greywater is via the branched drain system Ludwig developed. Branched systems use gravity and a specially fabricated pipe fitting that Ludwig calls a "flow splitter." Although these special fittings look very much like standard plumbing tees, they are not. The part is best conceptualized as an abrupt, though rounded, letter "Y." Often called a "double ell," they can be visualized as two standard, 90-degree plumbing-elbow parts connected back-to-back, and essentially with no back at all. As long as these parts remain level, flow splitters divide the quantity of greywater that drains through them precisely in half. Standard plumbing tees will inevitably send more of the greywater to one of the branches of the system and leave the other branch higher and much drier. Typically, flow splitters do not have this problem.

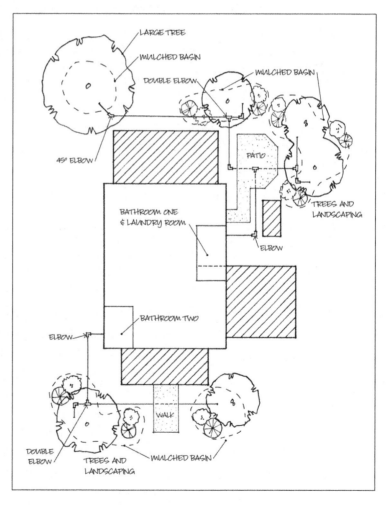

Harvest Your Grey

By using Ludwig's flow splitters, a properly designed and installed subsurface greywater irrigation system can convey predictable quantities of highly nutrient-rich water throughout a landscape to individual planting beds, shrubs, trees, or a compost pile. It's a wonderful way to harvest wastewater if your site, landscape design, and lifestyle can accommodate it.

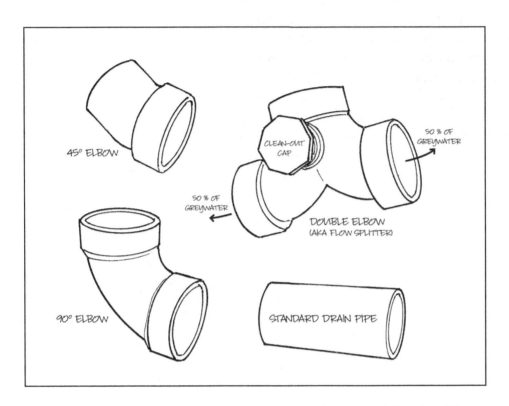

Ludwig's Flow Splitter and Associated Parts

One catch is that this valuable resource can only be diverted downhill. This means that your landscape design needs to be compatible with your home's plumbing fixtures, and this is not always the case. Even when a property's slope permits the use of greywater, harvesting it can sometimes force the landscape design in certain limiting directions.

Another drawback is that greywater must drop out of its distribution pipes into a mulched swale or leach-field infiltrator. In the case of the swale, the water immediately soaks into the ground via the mulch. In the case of the infiltrator, the water can percolate very close to the surface of the soil at the root zones of perennial plant material. Infiltrators cost more than mulched basins, but such basins are worth avoiding in any areas where toddlers might be playing or where people might wish to gather or walk across.

Another way to distribute greywater onto more than one place in your yard is the "twisted tee" method developed by water systems' designer Jennings of Earthwrights Designs. This system's main advantage is that it uses no specialized parts such as Ludwig's flow splitters. However, it requires a fair amount of trial, error, and hands-on tweaking because each standard plumbing tee needs to be turned to precisely redirect a desired percentage of water in the pipe. Making matters more difficult is that fact that low and variable pressures associated with greywater affect this percentage.

Ultimately, Jennings's system works on the same principles as Ludwig's. The former just makes sure his greywater collects for no more than 24 hours (less is best) in a surge tank so that when the water flows into his submulch greywater pipe, it comes out with a force strong enough to branch out through the various tees that he has situated just above the roots of certain plants. In addition to providing enough pressure for his system to function, the surge tank also works to cool down his washing-machine water before it heads out to his delicious golden currant bushes.

If your level of patience is high and your desire for precision is low, if you don't want to bother with purchasing any of Ludwig's flow splitters (which cost a little bit more than standard parts at your local hardware store), or if your site is perfectly suited for twisted tees, I'd try Richard Jennings's system—especially if you happen to be a hands-on, do-it-yourself type. Standard plumbing tees twisted at the precisely proper angle on a greywater distribution pipe serve as the 50/50 flow-splitting devices described above. In most situations monitoring and maintenance of these tees becomes more of a chore than monitoring Ludwig's flow splitters, but the initial cost is also significantly less for twisted-tee systems.

Please also note that laws regarding the uses of greywater vary significantly from governing authority to governing authority. Places in the nation with some of the more progressive greywater laws include New Mexico, Arizona, Texas, and certain parts of California. In residential situations,

New Mexico law allows for the unpermitted reuse of greywater on compost piles and landscapes as long as the distribution method complies with the following:

- Provides for overflow into a sewer or on-site wastewater treatment system
- Restricts access by humans, mosquitoes, pets, or "other vectors" into any storage tanks and piping associated with the system
- Dumps greywater outside of floodways and watercourses
- Releases water at least 5 feet above the local groundwater table
- Identifies any pressurized greywater conduit as "nonpotable"
- Ensures that greywater stays on site and does not infiltrate into neighboring properties
- Does not allow greywater to stagnate for any significant length of time
- Prevents greywater from being sprayed
- Harvests less than 250 gallons of greywater per day
- Complies with relevant local ordinances

These ten aspects of New Mexico's greywater law can really be boiled down to one commandment: Don't hoard your greywater; use it as immediately as possible in passive systems that are heavily mulched, appropriately contoured, and regularly monitored.

Construct Wetlands

Although very little of this book is designed to scare you, this chapter has a fearful quality to it. We are polluting our grandchildren's aquifers with our septic tanks and leach fields splattered all over the fruited plain. Did you know there's a reason that county ordinances prescribe a minimum number of feet that a well must be away from a septic system? It's that this distance prevents water-quality problems from occurring until at least a couple of decades after the permitting bureaucrat, the well-driller, the original septic guy, and the original homeowner are dead.

Based on how often we change homes, the average American septic tank probably has at least five or six owners during its lifetime. In addition to wondering how many homeowners even know about the importance of calling a septic service every two to four years to empty the tank, I also wonder about how many folks forget that inconvenient "no dumping hazardous chemicals down the drain" maxim. From sea to shining sea, septic systems are the second most cited cause of groundwater pollution—after chemical-liquid seepage.

Conventional "septic systems" are very simple. Wastewater dumps into the first of two compartments of an underground tank. Solids settle at the bottom of the first compartment and liquid wastes rise until they spill over a wall, called a baffle, into the second compartment. When the tank is at capacity, slightly less filthy water pours into a "leach field," which is typically composed of two trenches filled with gravel that are each also covered with a plastic infiltrator that allows water out while preventing roots from clogging up the gravel. Originally designed as systems that would be used temporarily— only until a municipal sewer system was made available—now after many decades of use these systems often allow waste to migrate toward the local water supply. So, the question becomes, "What kind of crappy water are we bequeathing to future generations?"

One alternative to the standard septic system is the kind of constructed wetlands that Natural Systems International (NSI) has pioneered. "Our systems feed plants and wildlife instead of polluting the groundwater below," explained engineer Erin English. Over a plate of native tacos one snowy day a few blocks down from her office, we were quite literally talking about amazing shit. "Constructed wetlands prevent subsurface wastelands by providing oases on the surface of the soil. This is not only beautiful, but it can also greatly increase local biodiversity and replenish a local aquifer with incredibly clean water."

Constructed wetlands run wastewater through a biological filter made up of plant roots, bugs of all sizes, some soil, and a fair amount of oxygen. According to the NSI website, www.natsys-inc.com, if designed, built, and maintained by competent people, you will eliminate "99.0

to 99.9 percent of all pathogens, including viruses" and significant amounts of the nasties that end up on the downhill side of any toilet. From total suspended solids and petroleum hydrocarbons to ionic and solid metals, "treatment wetlands," as NSI also calls them, will filter out about "40 to 80 percent of the total nitrogen in wastewater."

In addition to being able to significantly reduce the size of any leach field, wetlands produce irrigation-quality water at the far end of the system. "They can be very cost competitive on scales of eight to 12 homes," English said. "Smaller systems are more likely to be abandoned or forgotten, just like septic tanks often are, but in our work we like to work within appropriate economies of scale whenever possible."

NSI engineers projects all over the world, from thousand-home subdivisions in China to specialized residential projects just down the road from their Santa Fe headquarters. Schools and other government buildings and community centers are often ideal candidates for constructed wetlands because the need for a high-impact landscape can be as high as the large quantities of wastewater that such structures can produce. Often for institutions desiring LEED building certification, installing a constructed wetlands is by far the most cost-effective option because wastewater can be used to quickly establish shade trees on the south side of a structure (for the purpose of reducing the need for air conditioning) and evergreen trees on the north side of a building (for the purpose of insulating a building from its coldest exposure).

Ironically, wetlands can be especially cost-effective in high groundwater situations. Since they require a much more shallow profile and produce a much safer discharge than traditional septic systems, wetlands are one of the preferred methods of water harvesting in wet areas, although they work equally well in dry situations, too. The limiting weather factor can be cold temperatures, since the systems rely on the activities of living plants to clean the water. Because root systems slow down and can become inactive when soil temperatures dip deep below freezing, constructed wetlands need to be sized either to be able to do their job in a worst-case scenario of extended low temperatures or they need to be diverted to a conventional leach field during the winter months.

Harvest Your Waste

It would be impossible to describe all of the eco-features associated with the house that my late friend Mitchell Smith's company, Solarsmith, built for a client out at the end of Nine Mile Road somewhere in the middle of Santa Fe's urban sprawl. With its 400-square-foot array of photovoltaic panels, its 500 square feet of solar water-heating panels, its nearly 7,000 square feet of shaded and/or passively cooled space, its superinsulated skylights, its passive-solar radiant-heating system, and its sustainable forest-initiative-certified lumber, it was no big surprise that the home won the 2007 Grand Green Award for the most ecological home in the highest-priced category during Santa Fe's annual Parade of Homes.

Fortunately, Smith also spent some time thinking about the eco-home's water features. In conjunction with his team, Michael Nelsen from EcoScapes, LLC, and Jennings, Smith developed a plan for installing a couple of pumice wicks off the north-side entrance portals and then sending the rest of the megaroof's runoff into a 15,000-gallon underground cistern to the southeast. There was even a sign during the tour of the new home that that said "Future Vegetable Garden," and the well-mulched patch of soil was perfectly laid out between the kitchen and the carport.

The house was nothing short of miraculous from every aspect, but it was its wastewater system that made me realize how our water future is full of promise.

The house was nothing short of miraculous from every aspect, but it was its wastewater system that made me realize how our water future is full of promise. Topping off the property's list of eco-perks was an orchard and a couple of berry patches that were being watered by treated effluent from the home. Author William Irwin Thompson says that "when we have learned to recycle pollution into potent information, we will have passed over completely into the new cultural ecology." Taking pollution and turning it into brain food like berries makes me think this transition to a "new cultural ecology" may have unofficially begun.

Having perhaps the most appropriate appellation in all of this book's water-wise argot, "Sludgehammer" systems turn sewage into irrigation-quality water by means of crap-eating microbes and fast-moving bubbles that can be simply installed inside an existing septic tank. Bug populations build up and scale down their populations as needed. Typically, monitoring is necessary only once per year.

"These highly complex biological systems are extremely simple on a mechanical level," Jennings explained as he popped open the lid to his system. "Their true beauty though is that they're the cheapest systems around because the labor associated with these projects is not particularly elaborate."

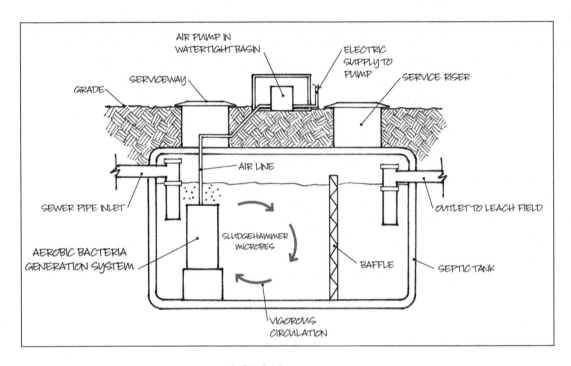

A Sludgehammer

Jennings also designed Sludgehammer systems whose treated water is pumped into toilets at a multiunit, mixed-use project called the Art Yard in Santa Fe's up and coming Railyard district, and he designed a different system at El Corazon, a 72-unit development a few blocks from the historic plaza. With these high-profile projects, Jennings is at the cutting edge of what needs to be done throughout much of the modern world.

There are other products that work like Sludgehammers, but they often require more technology and a particular tank or set of tanks that must be purchased. The distinguishing feature of a Sludgehammer is its ability to be easily installed in an existing septic tank. Now, harvesting your waste becomes an affordable option for anyone worried about polluting the aquifer under your leach field and anyone hoping to grow some percentage of one's own food.

As I hopped on my bike and turned back to the south, Smith's jewel of a house was immediately enchanting from the road. Rising slowly from behind a wide, organic-straw-mulched and native-grass-seeded knoll, the proud structure's smooth, rotund forms seemed, from a distance, to smolder like coals. In a polyvalent glow ranging from raspberry to melon to chocolate, its thick walls and puebloesque rooflines almost hummed in harmony in the noonday sun. The experience was spellbinding. Not surprisingly, Smith's house also won the home tour's Best Exterior Character Award. This kind of beauty-contest victory helps the sustainability movement as much as any Sludgehammer installation ever will. But, hey, we'll take converts wherever, whenever, and however we might chance to snag them. Whether people are attracted to the beauty or the functionality of sustainability is almost irrelevant.

Go Waterless

Although this book is not about conventional water conservation, I can't ignore two unconventional strategies that reduce the amount of water used when going either number two or number one. In fact, they eliminate the use of water for treating human feces altogether. The first one we we'll discuss is the composting toilet. The second is the waterless urinal.

The Western world's foremost authority on composting toilets is Britain's Joseph Jenkins. Author of the fun- and fact-filled book *Humanure,* Jenkins claims that properly composted human feces and urine can be used on vegetable crops, and he insists that water and excrement should be segregated as much as possible in order to save water as well as loads of valuable organic material.

Throughout the modern world there are two factors that prevent the use of composting toilets. One is that few of us want them. We prefer a loud swishing sound that takes our feces far, far away. The other factor is that regulators throughout much of the western world avoid the approval of human-poop composting. Typically these folks prefer no "creativity" whatsoever when it comes to changing any building code, especially the part of the Universal Plumbing Code that deals with human excrement.

Bottom line: unless an unusually proficient treatment facility is available, the flush toilet results in "wasting" two resources and creating a whole mess of unnecessary pollution in the process. For one person, on average, a five-gallon toilet flush contaminates each year about 13,000 gallons of fresh water to move around 165 gallons of body waste. Instead of creating vast quantities of resulting toxic sludge and building additional mega-treatment plants, our modern plumbing systems could include options for composting toilets that turn "humanure" into a perfectly good fertilizer.

Of course, a change in perspective this profound is not in the cards, at least not in the foreseeable future. Much more likely would be a scenario in which homes and communities across the land adopt the whole slew of other water harvesting techniques and strategies scattered through this book before we break up with the porcelain goddess that has been part of our heritage for so many generations. Believe me, I'm the same way. I actually have a great place for a composting toilet, but I would be surprised if my gradual-greening schedule gets around to including such a construction project for 10, 15, or maybe 20 years. Renovating my bathroom to fit a composting toilet is not a high priority in my life at the moment.

And right now, most places in the overdeveloped world won't let you install a composting toilet. Instead they force people to pollute local water tables with septic tanks and leach fields, or they make you pay every month to treat the sewage at some downstream facility. Perhaps some talented, zealous, and water-conscious lawyer could challenge local building regulations on the basis that the government does not have the right to take away valuable compost materials from its citizens, but anyone who tries this tact should be prepared to field a little hostility from the local plumbers' union and some friendly but change-phobic regulators.

Composting toilets can be divided into two species: one where the compost is created directly underneath the toilet and another where human excrement is taken away from the house (but not necessarily off the property) to be composted.

Prefabricated toilets that create compost directly underneath the toilet seat can be easily purchased online. Some use water, some don't. Some use electricity, some don't. As the technology gets more complex, the cost of these units rises.

Homemade in-house models can be very inexpensive, but they can require a significant amount of time to fully research, properly design, and efficiently build. In most cases the design and engineering aspects of such a project will be far easier with new construction, as opposed to bathroom remodels. The biggest difficulty with any kind of home retrofit would be having enough room for the requisite two or three chambers that every two to four years have to sit completely covered as they decompose and turn into compost (with the help of dry carbonic materials added after each use of the toilet).

Jenkins says his "bucket-removal system" uses one gallon of water for every 30 gallons flushed these days. For balance and efficiency, he recommends hauling two reasonably full lidded buckets at a time out to your compost pile, digging a slight depression in your pile, dumping your buckets into the pile, covering it back up, and then covering the pile with straw, weeds, or other organic mulch material. Then with a simple soap and water solution and a toilet brush, wash your buckets with half a gallon each.

All forms of composting toilets require more maintenance than flushing, but as Jenkins points out in *Humanure*, "When you use a composting toilet, you are getting *paid* for the small amount of effort you expend in recycling your organic material. Your payment is in the form of compost." By contrast in the conventional world of toiletry, he adds, the act of flushing symbolizes numerous payments that the toilet owner must make to water companies, electric utilities, and sewage-treatment facilities.

It's worth noting that Michael Reynolds of Earthship fame originally recommended composting toilets. Then his company decided to market a waterless poop-oriented oven called a solar toilet, and as result of their experience, in most cases now they simply suggest flush toilets are not the end of the world as long as the resultant blackwater is put to good use in the landscape.

Although our culture may not be ready in the first half of this century for a conversion to flush-free composting toilets, one technology that is gaining converts is the waterless urinal. There are two types of waterless urinals: a flush-free device that looks like a regular urinal and is relatively low tech and the tree in your own backyard.

Waterless urinals first came on the scene in the 1990s and have now been installed everywhere from Walt Disney World (where rainfall averages under 15 inches per year) to Alta, Utah (where a state record for annual precipitation was set at 108.54 inches back in 1983). At Alta, the six urinals installed save about 40,000 gallons of water per year.

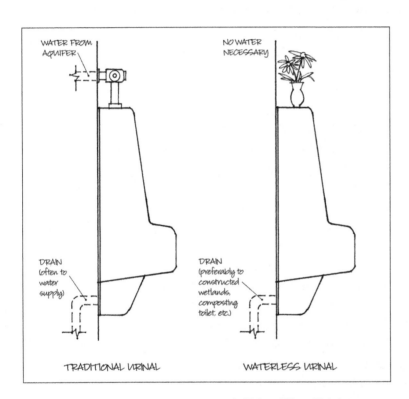

A Tale of Two Privies

My alma mater, St. John's College, installed waterless units not too long ago on its campus in Santa Fe. Although the Buildings and Grounds Department workers are not thrilled to service them, they save at least a gallon and a half of water per use, so they know they can't really make a big stink. "From a maintenance perspective," says custodial supervisor Glen Lopez "my men don't like to mess with them, but we do what we got to do."

According to one manufacturer's website, daily cleaning procedures are the same for flushless as they are with flushing urinals, and an article written by the Energy and Engineering Department of the Secretary of Defense stated, "After a balanced consideration, the waterless urinal seems to be a water conservation fixture who's [*sic*] time has come. It clearly reduces maintenance costs and may do so dramatically and immediately for some installations."

Relatively few complaints for waterless urinals seem to register over the Internet—especially when one considers the number of complaints associated with people like me who like to let it mellow. Most complaints are about odors that come from units that have been ignored for far too long.

Here's how these systems typically work: Waterless urinals connect to regular wastewater lines but eliminate water supply lines. Urine passes through a chemical barrier in a trap drain. The trap's biodegradable, recyclable seal blocks gasses in the sewer system from entering your bathroom. Your urine, being 99 percent liquid, runs down the drain as it normally would. The chemical barriers need to be either cleaned and/or changed from time to time, and this can be done during standard bathroom maintenance.

The great outdoors always remains an option for adventurous types who have privacy in mind and can appropriately and relatively quickly access large bushes and big trees.

5

Community Water Harvesting Opportunities

Starting today, we must pick ourselves up,
dust ourselves off, and begin again
the work of remaking America.
—President Obama's inaugural address,
January 20, 2009

Embrace Community

Just as I define water harvesting broadly, I paint the concept of "community" with a thick, wet, and far-reaching brush. Our human communities consist of the people whom we might directly affect with our actions. It's our families, friends, neighbors, coworkers, customers, and suppliers. Its people we talk to after church, during school, on our daily commutes, in our monthly book-club meetings, and wherever we end up having our share of reasonably meaningful conversations. Our community also includes people we never see. Maybe we influence them with a letter to the editor or through a Facebook post. Perhaps we have an effect by means of some random statement we make on a YouTube video or via a community-access TV channel that films public testimony at city hall.

In chapter 5, as in the rest of this book, you'll find colorful, caring characters designed to inspire rather than intone from on high, who thrive on ideas rather than dogma, and who seem to prefer interesting questions instead of easy answers. Most of them I know personally, and I feel very fortunate to be part of our loose and friendly community.

It's an exciting group. Our passions sometimes resemble my perception of our nation's first revolutionaries, who acted from gut instinct as much as from knowledge and thereby

Anyone can participate. Everyone has an essential role to play.

enabled the vital vision of change in their day. "Dry times are coming! Dry times are coming!" One can almost hear the voice of a reincarnated Paul Revere riding his bicycle through suburbia warning everyone he meets about the dangers of a fast-shrinking aquifer.

One important function of the community water harvesting part of this book is to remind us that anyone can participate. You don't have to be a gardener, a plumber, or even a landowner. The water harvesting revolution has already begun, and, yes, we are many. Yes, we are powerful. But we need you to participate, too. *We* are the community with the power to pass real solutions on to future generations as the water crisis intensifies. Leo Tolstoy was right when he suggested in *War and Peace* that it is regular people who really matter when it comes to change—as much or more than our world leaders. Everyone has an essential role to play, and often it is those who are given the least power who have the most liberty to try new things.

Like democracy, the water harvesting revolution is not a spectator sport. Each requires significant human participation. I hope you find a way to join in this movement toward concrete change. What follows are some examples designed to excite you into greening your professional, social, and/or political life by at least a little bit—but maybe, hopefully, by a lot. Finally, please remember that these sections are much less about specific examples to follow than they are about general strategies that you might want to consider for the purpose of restructuring them within your own unique set of circumstances. Just as we can find extremely different natural environments among adjacent bioregions, our communities vary widely from place to neighboring place.

Develop Green Neighborhoods

Out of a basic instinct for self-preservation, it is in the development industry's best interest to lead the charge toward a sustainable water future. Clearly we are witnessing new trends in design that are in keeping with principles of sustainability, healthy environments, and smart water and energy use. Old-world real estate and development practices are coming under fire as costs and property taxes mushroom. In a new world of real estate, freed from old verities of man-over-nature, the development of communities will align with natural forces, elements of the physical environment such as storm water, wind, wildlife, sunlight, and any change in grade inherent in the given terrain. Cultural forces, such as vehicles and roads, bikes, pedestrians, right of ways, and utility lines, will be rethought and old constraints thrown off. Just as individual landowners can increase a property's value by water harvesting in the ways that I have described, developers will begin to improve their real estate investments by designing communities that work effectively with these greener influences.

Natural forces too are best understood as potential resources. Too often developers think of runoff as a waste product that should be directed as quickly as possible away from homes and roads. Imagine how much more beautiful and comfortable our subdivisions would be if runoff were given a chance to percolate into the local soil. Imagine if storm water were directed to windbreaks and shade trees, which in turn would provide protective microclimates for healthy native grasses and flowers.

Similarly, cultural forces are best understood as elements that enhance, rather than hinder, the quality of life in our communities. Too often we demand too much space for our motor vehicles and not enough access for alternative forms of transportation. Imagine how much happier and healthier we would be if we had plenty of bike paths, safe sidewalks, and easy access to reliable public transportation. Perhaps the solution to high heating bills could be an increased number of passive solar homes packed next to each other with shared (and well-soundproofed) walls. It might be that the best answer to a rising crime rate is more "eyes on the street," the old-fashioned way of people looking out for their neighbors.

One developer who understood these concepts many years ago is a man named Don Altshuler, who, along with a small group of other like-minded folks, founded the Commons on the Alameda in Santa Fe back in 1992. For several years before they could move in, Don and his cohousing project rented an office next to the sustainable-communities nonprofit where I worked. He was a great neighbor then, and after so many years of success, it is not surprising that the Commons turned into the most desirable subdivision in the bioregion.

The Commons on the Alameda

Like most other cohousing communities, people have their own individual kitchen / dining areas and their own small gardens, but "commoners" also share a number of amenities. For the 28 families who live there as well as their friends and relatives, these include a common kitchen and dining room, a plaza with a fountain and plenty of shade, a couple of guest rooms, regular childcare, play structures, a basketball hoop, a chicken coop, an orchard, an organic garden, and a beautiful arroyo that only a decade ago looked like a dismal failure from an erosion-control standpoint.

Don and his group were committed to creating a community, not just another unappealing subdivision that would simply create more separation among people. In the typical subdivision, kids have their own play structures and basketball hoops, and adults, if they have a garden at all, benefit from none of the shared information that comes with a community garden. The lists of benefits of such places goes on and on, but now, with the reality of the debt crunch settling in on American soil, the economic benefits will become the most appealing of all because if a cohousing community is done correctly, it should be a relatively inexpensive place to call home.

In the case of the Commons, medium-sized houses on very small lots have sold for nearly half a million dollars, so whether the goal is high-end pricing or affordable housing, it shouldn't take the development industry much more time to figure out how the wave of future development will include creative endeavors like cohousing and the larger movement known as "new urbanism."

A recent example of the new-urbanist model now exists at Oshara Village just down the road from the Santa Fe Community College on the south side of town, where most future development will occur. "All of the water in Oshara will be returned to each home and park for irrigation," said Alan Hoffman, one of the principals of the project, while relaxing under an EZ-Up tent at its inaugural community event. "Between reclamation and water-conserving fixtures, a family in Oshara could use as little as half the potable water of a conventional home in a conventional subdivision development."

Hoffman, a realtor and former builder of solar homes, has worked extensively with new-urbanist architect Andres Duany. In addition to caring about the water resources of a community, Duany, the author of several books on urban planning, is dedicated to the concept of providing a convenient, walkable center for every development in the new-urbanist model. Developments should be "mixed use," he says on a website video at www.osharavillage.com. "There should be places to live, places to work, places to shop, places to go to school and places to gather." In a world of rapidly increasing fuel prices, these kinds of communities represent the ecological and socially supportive villages of the future, not only because they are more comfortable than urban sprawl, but also because they are more efficient and, therefore, less expensive to inhabit.

These new approaches to development need and are, more and more, finding visionaries at the city, county, and state levels who see every square foot of every rooftop as a resource. Architects, engineers, builders, plumbers, and landscapers must begin to see sinks, showers, *canales*, and downspouts as ways of increasing a property's value. Conscious realtors and lenders will also be needed to promote the benefits of these new approaches. And, finally, as more and more landowners begin to demand the alternatives that exist, the market will then be driven toward the respect for water that our species now requires.

Increasingly, permitting agencies will see the benefits of encouraging such developments, while buyers will be attracted not only to the minioases that can be created by these developments but also by the cost savings that come with cohousing and new urbanism.

Invest in the Rain

In this age of dwindling water resources (and scarce financial ones), if communities are ever going to begin to harvest precipitation on the scale necessary for the continuance of the conveniences of modernity, both the public and private sectors of our economy will no doubt need to take on leadership roles. Unfortunately, credit crunches notwithstanding, lenders, investors, government agencies, and business types can be a relatively cautious breed when it comes to getting enthused about such long-term returns on their investments. Even so, signs of change are beginning to sprout.

One advantage is that swales, cisterns, and Sludgehammers are not real estate, so they represent a much smaller risk in terms of dollars that need to be invested. Active and wastewater harvesting projects are more akin to the scale of car financing. Depending on their size, their efficiency, and the number of desired "extras," active water harvesting systems and wastewater-treatment methods range in price from "used and abused junker" to "new and souped-up Toyota Prius." In a few cases, residential cistern systems can compare to the brontosaurus-sized price tag of a newly minted Hummer or some special-edition stretched Escalade.

Another advantage is that unlike an old vehicle with hundreds of thousands of miles and certainly unlike a wasteful Humvee, cisterns and water-recycling systems will typically provide an instant asset to any property and for any household. As water harvesting has a positive effect on the landscape associated with it, the value of the particular technique, system, or method appreciates over time as it simultaneously reduces the typical family's water bill. Fortunately, even in this difficult environment, some lenders and investors are ready and willing to infuse cash into these types of projects, and if we organize effectively we will also see local, state, and federal governments start to play an important role in the water harvesting revolution.

At the Permaculture Credit Union (PCU) in Santa Fe, you get an automatic discount on your loan if your project includes water harvesting (or has other green qualities). It's called PCU's "sustainability discount," and it's not much more than a phone call or an e-mail away for those who qualify.

There are only two things that you have to do to become a member of the PCU: give them five dollars and say you believe in the ethics and principles of permaculture. And, if you are uncomfortable taking such an informal vow, you can even simply become a member of any recognized permaculture organization and you're in.

Thanks to more than three years of volunteer work by a group of dedicated volunteers, about a decade ago the PCU became the first new credit union chartered by the state of New Mexico in 30 years. Like all credit unions and banks, PCU provides dividend-bearing savings accounts and competitive CDs, both insured by Uncle Sam. Unlike banks, however, credit unions are nonprofit institutions that have members who can participate in a one-person, one-vote process to control the use of the institution's money.

The PCU defines "sustainable" much more comprehensively than people might think. If you wanted to borrow money for a wasteful lawn-sprinkler system or if you wanted to buy mineral rights for the purpose of extracting gas and oil, getting a sustainability discount would take some serious smooth talking. "But the argument could certainly be made," says President/CEO Donald Sarich, "that a big gas-guzzling truck could be a vehicle *for* sustainability. What if someone were to use their Dodge 3500 to pick up and deliver locally cured compost?"

The people at the Permaculture Credit Union discovered a long time ago that determining what is and what isn't sustainable would become a counterproductive waste of time. "I don't mind being the one who gets to decide where our members' money is invested," Sarich said flatly before breaking into a thin smile, "but I sure wouldn't want to be the guy drawing strict lines between what is and what is not sustainable."

The PCU's sustainability discount dangles real money in front of anyone who might need extra encouragement when it

"There are profitable opportunities out there for every eco-investor."

comes to making the world a better place. But it does this in part because it also makes good business sense to do so. If a member wants to buy an energy-efficient appliance, a hybrid vehicle, or some insulation for a home, the appliance, the vehicle, and the insulation will each reduce another monthly expenditure, and this increases the likelihood that their members will be likely to pay. A decade after the PCU's inception, Sarich said that the member-owned institution was having its best year yet.

The world of asset management also has a growing number of people who care about more than just the traditional kind of green. Natural Investment Services is one of many eco-advisers that focus on values-based investing. "Although our primary mission is to provide our clients with the returns they desire, we attract a clientele that also understands that financial decisions have real consequences in the world," explained my friend Michael Kramer, a former Santa Fean who is now a managing partner and the director of social research at NIS. "At the same time, in the face of many critical social and environmental challenges, we see vast opportunities in our global economic system. Like it or not, money is one of the most powerful forces for determining human destiny, and a tsunami of responsible investment is on the horizon."

For water-responsible investments, Kramer, from his desk in Hawaii, suggested the Sustainable Water Fund offered by Sustainable Asset Management (SAM), a newly available mutual fund that has only recently been offered in the United States. He also warned, though, that "if you look at the underlying holdings of the Sustainable Water Fund, there will be a number of companies that the purist might want to avoid. Still, it's the most forward-thinking mutual fund when it comes to water issues."

Completely independent of my conversation with Kramer, *Green Money Journal* publisher Cliff Feigenbaum concurred. "There are profitable opportunities out there for every eco-investor," he told me as we walked through the parking lot at the Santa Fe Farmers' Market. "With the population of the world tripling and per capita water consumption doubling since World War II, UNESCO is predicting a major drop in water resources in less than a generation. A fund like SAM's Sustainable Water Fund would be a great place to start for anyone wanting to make a difference when it comes to water issues."

At the government level, things are also beginning to shift in a positive direction. For years incentives such as rebates for cistern systems have been discussed at meetings of the City of Santa Fe's Water Conservation Committee, and after a successful low-flow toilet retrofit program that replaced most of the city's old, wasteful toilets with highly efficient models, "the city is in a pretty good place to revamp that program in a way that promotes cistern systems," according to the committee's vice chair, Melissa McDonald. "The details do need to be worked out, but it's an idea that ought to be applied in the real world." Part of the problem, said McDonald after one meeting of the committee, is that specific guidelines for cistern installation in our climate have not existed up until now. "Fortunately the city and county are on the verge of completing a set of guidelines that could help any rebate program function in a fair and effective manner. Pretty soon, it may just be a matter of getting enough public support to fund the program."

At even higher levels of governance, the New Mexico Office of the State Engineer (OSE) and the Environmental Protection Agency (EPA) have both been doing their part. The EPA recently came out with its own set of guidelines for cisterns, and the Water Use and Conservation Bureau of the OSE has just published a free online how-to manual called *Roof-Reliant Landscaping*.

For the OSE's conservation bureau chief, John Longworth, one significant concern in publishing such a manual is that people will design and install a cistern and suddenly think that this entitles them to use *more* water. "We've got to nip this one in the bud," he told me when we were first getting started on writing the "cisterns meet xeriscapes" manual. "People have to realize that if their water harvesting systems do not work within a rational water budget and a feasible landscape plan, they remain part of the problem." By making *Roof-Reliant Landscaping* about appropriate landscape design and planning as much as making it about cistern systems, "we were able to fulfill the goal of providing useful information to a public in desperate need of direction when it comes to the efficient and productive use of the precipitation that falls on their rooftops," he said.

If you have read the first parts of this book, the abundant business and professional opportunities associated with water harvesting should be obvious. In the coming decades in many regions of the world, there will be a predictable uptick in the number of homes with swales, check dams, cisterns, pump houses, greywater systems, Sludgehammers, composting toilets, and all of the other techniques, systems, and methods that we have discussed. Plumbers, contractors, engineers, developers, and people managers who can fill the niches created by this uptick will, by nature, be in greater demand, as will ecological landscape-consultation, -design, and -installation firms.

In any given community, the smart entrepreneurs and tradespeople will want to get together to collaborate with each other. For the water harvesting industry nationally, the American Rainwater Catchment Systems Association provides chat rooms, list serves, seminars, conferences, and many other means for learning and networking. On the local level, you might join or start a group like the one Reese Baker and Richard Jennings started in Santa Fe: the Semi-Arid Café is a regular, though informal, gathering of local water harvesters. On a monthly basis, the group comes together, typically in someone's backyard (in the summer) or a local restaurant (in the winter), to share their experiences, knowledge, and new ideas.

One of the people I met through the cafe is Doug Pushard, a Santa Fean who runs www.harvestH2O.com. Dedicated to the advancement of sustainable water-management practices for individuals, families, communities, and businesses, the site is an expansive clearinghouse of information about water harvesting. With scores of water-related articles and countless links to the tools and materials that water harvesters need, Pushard is a great example of a water harvester who is using a business investment in the newest of media to teach people about water harvesting.

According to Pushard, his website is visited by 20,000 people every month. In its tenth year of operation he was particularly happy because, he believed, his site was finally poised to make a profit. A software engineer by day, Pushard had been burning late-night oil with his website posting hundreds of newsworthy items and how-to information about rainwater harvesting in all its forms. "The website is a lot more alive than it used to be. After working for years to improve it," Pushard said during a meeting of the Semi-Arid Cafe, "we're finally at a point where people can do detailed research on specific topics that they care deeply about. I'll never make a ton of money doing this stuff, but soon I'll be able to focus fully on precipitation, instead of software, and this will represent an elevated step for me in my life."

In terms of investment, note that none of these examples were about gaining instant profit. With plenty of patience and realistic aspirations, these rainmakers (from the lender and the two investors to the two political types to the entrepreneur) were looking toward long-term success, not short-term gain. The sprout is up and it's strong and hopeful, but don't look for it to take off in the near future. It has many more years to build its root systems before taking over for good.

Teach Kids

In addition to Pushard's web-based way of sharing information and the traditional meeting style of conferring with colleagues, traditional teachers who have knowledge and/or experience about the solutions to our water problems will also become increasingly popular at every level of schooling, from prekindergarten through postdoctorate research. Farmer-educators in particular like Patty Pantano, who founded Camino de Paz School in Santa Cruz, New Mexico, with her husband, Greg Nussbaum, make water an important part of their educational institution's curriculum. The farm-based Montessori school's students focus on one of three themes over the course of their three-year experience in middle school.

"One year the theme for all of our seventh, eighth, and ninth graders will be energy; the next it's living things; and, finally, during the third year of this cycle we focus on water," Pantano told me over an egg-salad sandwich courtesy of the school's happy, healthy flock of chickens. "We go through water conservation, water tables, water testing, watersheds, water rights, the history of humanity's use of water, water harvesting, water treatment, water colors.... The list goes on and on. A lot of it fits very well within our math and science programs, but it often quickly spills over into each and every one of the humanities."

Mark Duran is another teacher worthy of note. In his environmental studies elective at Santa Fe High School, Duran has been able to pass on what he's learned from his studies of holistic management. Over several thick strips of cow meat and a local beer, Duran briskly ruminated on the importance of ruminants. "If you want to be a vegetarian, fine, but cows aren't the problem. The problem is one of management. If we took all of the hoofed animals out of modern feedlots and allowed them to graze in a way that mimicked the herds of wild buffalo that lived here for millennia, we would have a much healthier economy, a much more stable ecology, and a far more efficient and productive society."

Duran pointed out that in "brittle" environments, places that have at least one significant dry season per year, grazing herds, more than anything else, would most rapidly bring back topsoil. "As long as herds could be moved slowly when the grass is growing slowly and quickly when the grass is growing quickly—that's critical," he emphasized. "If done right, holistic management recharges aquifers, builds soils, increases biodiversity, provides incomes in rural areas, and produces a high-quality protein for human consumption. It's a winner, and my kids see that."

Teaching high school students is a challenge, Duran went on, swinging back his long ponytail, but "the biggest impediment to understanding these concepts is the amount of stuff that the kids think they already know. Every year there'll be some kids who start out claiming all cows are Earth killers, but by the end of the year I can usually get them to admit that grazing pressures are necessary for the health of any arid-land savannah. Although many of their parents may never get it, it's nice to know that we can make progress, steady progress, with the next generation."

Although it can be difficult at times, teaching middle and high school students is also one of the most rewarding experiences that I have ever had. Granted, I'm a lot like my Aunt Marjorie when it comes to infants: she loves them because "they are so adorable, and when they cry someone comes and takes them away." Similarly, I'm a big fan of school kids whenever I'm expounding about things I think I know, while they ask their questions, and especially when I get to witness their off-the-charts creativity, but, in the interest of full disclosure, I haven't been paid to control a group of kids since I was one of them, as a 17-year-old camp counselor. Still, thanks to my perennial gigs teaching everything from mulch and swales to cisterns and greywater, I know teacher pay is much too low, and I know that some days just suck, but ultimately middle and high school kids are the most inspiring age-demographic to be around.

Thanks to a Santa Fe–based group called River Source, over the last four years I've participated in a panel of water-oriented professionals who get to interact with four or five classes from as many New Mexico schools. It's an event called "Youth for a Secure Water Future," and every year I encounter a new, awesome group of high schoolers. They realize on a visceral level that our watersheds have been squandered by the grown-ups who are constantly telling *them* what to do. But even so, they are willing to look past these faults (for one thing, they're used to them) and get down to the business of creating meaningful solutions.

Especially when they hear about water harvesting, their eyes perk up. Since before they were born, container recycling (bottles, cans, jars, boxes, and bags) has been the societal norm, but when they hear that they can recycle the ultimate ingredient of life, they instantly relate to the authenticity of the goal. Traditional recycling to many of them has always been on the superficial plane of "packaging," but when they begin to understand that water can be recycled, you can almost see a fire of hope illuminate their faces.

Teach Adults

One of my favorite local educational institutions is the Permaculture Institute run by Scott and Arina Pittman. It's based out of their permaculture demonstration farm situated on a small promontory overlooking the Pojoaque River, about 15 minutes north of Santa Fe. In addition to the overflowing organic garden, the bountiful orchard, and the flocks of chickens and guinea hens that you would expect to see at a place like this, their "Lots of Life in One Place" project also features a *chinampas,* a lush, serpentine, chest-deep waterway created out of an unproductive marshland that been overrun with invasive and exotic species.

"We've seen a tremendous increase in otters, birds, insects, and aquatic life—everything from turtles to fish," said Arina as she took me on a tour of her happy homestead.

"We've got two types of horned owls, a number of raptors, several kingfishers, an occasional blue heron, and a neighbor has counted 12 types of ducks," she went on. "The only time we've ever had an insect problem is when we let the population of our flock of guinea hens lapse. But within a few weeks of hatching a new bunch of young chicks, everything was quickly back in balance."

The Pittmans teach about half of the institute's classes themselves, and the remaining half is taught by specialists in various fields, from beekeeping with longtime apiculturalist Les Crowder to orchard keeping with high-alpine fruit-tree guru Gordon Tooley. Together the Pittmans have been teaching permaculture for a combined 35 years.

"We have students who are activists, design professionals, policy makers, landowners, backyard gardeners, healthcare professionals, and people of creative professions who find permaculture expands not only their creativity, but also their entire worldview. Some people say, 'If you positively affect the life of one person you've fulfilled your destiny.'" Arina continued, "What I find is that teaching these classes makes me feel truly blessed because we're affecting the lives of so many people."

One message that both Pittmans like to relay about Lots of Life in One Place is the philosophy behind the name of their *ranchito*. During her travels around the world as an anthropologist, a mutual friend, the late Fiz Harwood, kept asking the people of Polynesia to describe the divine force that they worshiped, and the people would respond, "It's in fruit and butterflies," she said, "It's in children and in the fish." The concept of the divine for these ancient cultures grew out of the notion of abundance. "Fiz explained that divinity is present in all life but most especially wherever you have lots of life in one place," Arina grinned into the eyes of her newborn son. "Hence, the name."

Teaching doesn't have to occur within the confines of man-made institutions either. As I have had the pleasure of remembering, the best setting for learning is often Nature herself. In the summer of 2009, I had the pleasure of walking along one of the last tributaries of the Chama River before it joins the Rio Grande just north of Espanola, New Mexico. Led by my friend Joel Glanzberg of Regenesis Group, five water harvesting colleagues and I tiptoed in complete awe behind Joel as we came across centuries-old check dams still holding back arroyos, thick river-rock mulches apparently hauled up the sides of steep ridges, and seemingly endless on-contour swales made of black, volcanic cobbles.

"They basically planted all over the watershed because you can never predict where ample rainfall will occur," Glanzberg explained. "And they would also plant right along the stream banks. Sometimes their crops in one place would fail from lack of rain, but that might mean that other fields thrived. The idea was that the tribe would more or less split up in the spring and get together once in a while to share the bounty."

Like Socrates leading Phaedrus, Alcibiades, and the gang through dialogues of the mind, Glanzberg directed us down one ridge into the shade of hundreds of cottonwood trees into a kind of observational dialectic. In the middle of a tight valley, he pointed. "See how the high-water mark is over our heads right here?" he asked. "There can be plenty of water in a desert. We just need to learn how to use it." As one of the permaculture movement's great teachers, Glanzberg loves to share his knowledge about indigenous sustainability. Fortunately, he has plenty of it. Who is the Joel Glanzberg of your watershed? I hope you seek him or her out as soon as you can.

Become an Expert

I became a "recognized expert" in the water harvesting field quite by accident. In early 1992, I was working with SEED (Source for Ecological and Economic Development) and we were going to build an ecological community close to Santa Fe complete with every component of sustainability, from solar panels to cisterns. We were going to be *the* model community and if we hadn't suddenly lost our principal source of funding and if I hadn't suddenly been laid off, our model community would already be a couple of decades old by now. So it goes and yet, as is so often the case, when a door closes, a window opens. I climbed out and looked around.

The day after getting the news about my no-longer-existing job, I came across a newspaper ad for a permaculture design course with Bill Mollison. I'd heard about permaculture

As we've simplified, we've become more productive, less stressed, and increasingly known as "experts" in our chosen field.

and it had sounded exactly like what the environmental movement at that time needed, an encompassing production-based philosophy. I signed up for the class and loved it. A couple of weeks later, I ended up getting a job with one of the teachers, Derk Loeks, a shaman of passive water harvesting who owned a small permacultural landscaping company.

A year later, my boss decided he longer wanted to be a boss, and after an extremely friendly takeover of three of his projects, I was in deep and the business was growing. Fast forward a dozen more years and Santa Fe Permaculture, Inc., has a payroll of over 20 weekly paychecks. Meanwhile, my wife, Melissa, and I, who had been running the company full time, thought it good timing to raise a family. Then I happened to have two meetings that changed the course of my life. In one I was asked to write the book you are reading now. In the other, I was invited to submit a proposal for writing *Roof-Reliant Landscaping* for the State of New Mexico.

For years I had been producing a monthly column that appeared in the real estate magazine of our local newspaper. The column, *Permaculture in Practice,* had been an important part of Santa Fe Permaculture's marketing and my self-discovery: write a regular column for a target market and people will think you know what you are talking about (they did) and eventually you will know a lot (I'm getting there). Today, our firm focuses more on permaculture landscape design rather than on the grit of landscape installation. I'm balancing time now for writing, design, and being a caring husband and father.

As we've simplified, we've become more productive, less stressed, and increasingly known as "experts" in our chosen field, even though the more we learn, the more we realize we have much more that we should study. The field of sustainability is still new, so it's easy for people to think of us as experts when what we are is pioneers, at times working as much on instinct and intuition as with knowledge and science.

In addition to books like this one, the field of sustainability, permaculture, and water harvesting continues to attract new insights and key teachings from writers like Brad Lancaster. A longtime friend and colleague, Lancaster is an Arizona-based water harvester and permaculturalist. His current project is a three-volume set called *Rainwater Harvesting for Drylands and Beyond.* Filled with hundreds of illustrations and plenty of resources, Lancaster's first two encyclopedic volumes are front-of-field in promoting the effective use of precipitation all over the world.

Lancaster describes a wide variety of water harvesting systems. Chapter 1 of volume 1 starts with an inspiring vignette called "The Man Who Farms Water." Lancaster tells the story of Zephaniah Phiri Maseko, who over a 30-year period raised a family of 10 on a 7.4-acre piece of land in a region that often gets less than 12 inches of rain per year. With no money for a well, Mr. Phiri installed check dams, swales, two cisterns, and a host of other structures that slowed the flow of runoff water through his land. Over time, water (which would otherwise have created soil erosion) established orchards, vegetable crops, sugarcane, a dense banana grove, and enough drinking water to sustain chickens, turkeys, cattle, goats, and his family.

The chapter ends with an in-depth look at Lancaster's eight principles and three ethics of water harvesting. In subsequent chapters, he applies these principles and ethics to topics ranging from how to estimate your property's water needs to how to develop an integrated, water-conscious landscape design for your property. Along the way, he concludes each chapter with an example that puts these principles and ethics to work.

Having researched a significant number of water harvesting texts, I can assure you that there are few books that provide this kind of essential information in such a readable and well-illustrated format. As water resources become increasingly scarce, Lancaster's *Rainwater Harvesting for Drylands and Beyond* volumes will prove to be among the seminal books of our time.

Lancaster's story is pretty similar to mine. He worked for an ecological nonprofit, Native Seed Search in Tucson, Arizona, until his landscape-design business took off (at about the same time mine did). One day he realized there were a lot of water harvesting books out there geared for people in the developing world yet not one text for mainstream Americans, so he decided to start writing. And as much as writing and teaching are very rewarding to Lancaster and me, we both get incredibly invigorated by straightforward, down-to-earth ditch digging—especially if it's for a great cause like water harvesting.

One of Lancaster's most exciting projects is one that

captures runoff from suburban streets. With simple curb cuts and some sensible grading, Lancaster creates eddies in planting beds between sidewalks and streets. In a few years, neighborhood residents get shade and food crops where they used to get dust and glare. It's pretty simple stuff, but somebody's got to do it.

He and I also have a close-to-identical perspective when it comes to the "recognized expert" concept. "Anyone can become an expert," Lancaster told me through his mobile phone on his way to another speaking engagement. "They just have to do a few of these things in their own life, on their own pieces of property, or they just need to become part of a couple of larger group projects. Then they will learn what's best for their environment, and they'll be able to speak from real experience while always comprehending how every strategy needs to be handled in a slightly different way because factors in the environment change with every new site."

Support Water Justice

One career calling in which you can expect to make less money than most while you're on the front lines also happens to be among the most exciting and fulfilling. Consider being an advocate for water justice or, as many are now called, a "water warrior." In *Blue Covenant: The Global Water Crisis and the Coming Battle for the Right to Water*, activist, author, filmmaker, and United Nations water envoy Maude Barlow provides more than a dozen stories of Bolivian shoemakers like Oscar Oliveras who step up to stop corporate-giant Bechtel from owning his community's water, or of Italian visionary academics like Riccardo Petrella who spearhead such groups as the "International Committee for a Global Water Contract."

Activism for "water justice" is on the rise, and multinational corporations are preparing vigorously for political conflicts on every continent as a result. As Barlow says at the outset of her chapter called "The Water Warriors Fight Back," a mighty contest has grown between those (usually powerful) forces and institutions that see water as a commodity, to be put on the open market and sold to the highest bidder, and those who see water as a public trust, a common heritage of people and nature and a fundamental human right.

In some cases the fight is intense and dangerous, in

other places merely intense. It takes a certain type of person to be so caring and also so brave, but most of the time water warriors simply have no choice: they must become courageous, or they must die of thirst while watching friends and family suffer the same fate.

Water warriors remind me of the patriots of the American Revolution, with one or two reciprocal differences. Both must defend their lives and livelihoods, but instead of fighting an old monarch for a new way of life, water warriors battle new forces of corporate tyranny in order to protect their ancient ways. An influence from abroad charges a fortune for a local transaction. Instead of tea and taxation, a water warrior's job is about survival. No longer fighting the British Empire and its East India Company, these freedom fighters must wage war against more powerful and trustworthy-sounding entities, such as the World Bank, which has funded the corporate takeover of indigenous watersheds for decades.

When it comes to having a profound impact on your ability to have a positive effect on the looming water crisis, what you do for a living can contribute from eight to 10 hours a day (or more in the case of most entrepreneurs and water warriors), let alone the 10 minutes that we have been shooting for with gradual greening. Especially in these uncertain economic times, as more and more people discover that college loan monies have dried up, jobs have been shipped overseas or just eliminated, retirement accounts are depleted, and homes are worth less than ever, it may be time to set your sights on a new career. Who knows? With Wall Street looking a little less glamorous these days, maybe soon the brightest kids will get back into community organizing, which is, after food production, the second most practical of all careers.

Water justice, as an issue in New Mexico, predates the American Revolution by a century or two, and longtime problems continue to this day. As of this writing up and down the Rocky Mountains there's a high-stakes political poker game going on. Oil and gas firms are trying to use ancient laws to get at the fossil fuels near our aquifers and surface-water supplies. In Santa Fe, the Houston-based corporation, Tecton, has the 1872 Mining Law as its ace in the hole. Its king is our society's voracious appetite for energy. Its queen, a queen of diamonds, is the glittering allure of money that talks, and the oil/gas/

minerals industry has its share of money that's lubricating and paving the way for the Halliburtons and Mobile Exxons of the West and the world.

Although the deck seems stacked, there are many more cards to be played. Overflow crowds at public meetings in the Southwest and other regions are growing, accompanied by public interest and outrage, a barrage of e-mails, letters, and phone calls to public officials at every level. Letters to editors on drilling issues, e-mail campaigns and helpful websites, and efforts by well-known environmentalists like Robert Redford have all combined to mount successful defenses aimed at protecting much of our remaining surface water, groundwater, and lands from being exploited and/or polluted by Big Oil.

In addition, the people of northern New Mexico have had recent success in protecting the Valle Vidal, a 100,000-acre piece of national forest. Home to roaming buffalo, wild turkeys, black bears, mule deer, mountain lions, and bald eagles, the Valle Vidal lolls deep in the heart of the southeastern branch of the Rocky Mountains, the Sangre de Cristos. While large populations of reptiles, amphibians, and fish pulse through this stream-filled wilderness, New Mexico's national treasure sports the state's largest elk herd, an endangered species of trout, and some of the most beatific camping you'll ever find. A troop of several thousand Boy Scouts and myriad hunters, anglers, hikers, and bird-watchers bring over $5 million to the local economy every year—and much more to the state's economy in terms of the residual effects associated with having a reputation for possessing pristine wilderness.

Several years ago, the "Valley of Abundant Life," as the Jicarilla Apache Indians called the Valle Vidal, was being threatened by another Houston-based energy leviathan that wanted to pump methane out from under coal-bedded aquifers. In "coal-bed methane development," many of the Valle Vidal's age-old aquifers would have been pumped dry. This would have meant an incredible waste of our most precious resource, a serious reduction in surface-water supplies (which are directly related to neighboring aquifer levels), and dangerous downstream water pollution.

Fortunately, enough people stood up and told former senator Pete Domenici to stand down or they would stand up even higher and create more problems than whatever the payoff was worth. With the help of current senator Jeff Bingaman and voices of advocates such as Stewart Udall, former secretary of the interior, a large part of the Valle Vidal is now officially protected by the federal government.

Harvest Democracy

What will happen to the nearby Galisteo Basin and so many other natural treasures remains to be seen. Fortunately, groups like River Source and WildEarth Guardians that focus on protecting riparian habitat over wide regions are working hard to organize people against destructive efforts like Tecton's "hydraulic fracturing process," which would certainly destroy many of our freshwater reserves. Meanwhile, groups like Earthworks Institute and Drilling Santa Fe, which concentrate on the particular affected bioregion, are joining forces with

People who stand up to defend a local watershed are the water warriors of any community.

more than 50 other local groups in what should be understood as nothing less than a struggle to defend our homes, our land, and our livelihoods.

People who stand up to defend a local watershed are the water warriors of any community. They aren't making much money doing what they do, and going toe-to-toe with major corporate interests can be intimidating, but I'd bet they can rest well at night knowing that they are doing as much as they can to preserve that which sustains us all. They see a garden of delight in every stream and are compelled to protect it forever.

Build Consensus for New Ideas

My Dad has always been big into changing the world for the better. In fact, my earliest philosophical memory is of his telling me that "the purpose of life is to make the world a better place than it was when you entered it." He also likes to say that "there are two ways to change the world. You can change it from the inside, or you can change it from the outside, and making change from the inside is far more effective."

As is often the case with fathers and sons, we disagree on a few things. He prefers the challenge of attempting serious reform from within, while I have always felt more comfortable as an outsider. Sometimes it's not even clear which side your power is coming from, but fortunately, if the goals are to build consensus and move new ideas forward, then either way can work well. Running for office is a great example of this phenomenon because, win or lose, the ideas you put forth have an inherent power all their own.

I met my wife, Melissa McDonald, at a local Green Party meeting in 1992. Most of my college friends had already trickled out of town in search of a bigger phone book, and I'd been going to the weekly shindigs as much for the social aspects as for the political activity. At first I had thought Melissa was one of those beautiful blowhards always at the front of a volunteer meeting saying, "Do this, this, this, and this." But unlike so many other folks who disappear when it's show time, Melissa followed through and was always there for the logistical work of organizing both people and things. She could not only get everyone to make their phone calls and show up at events, but she could also easily materialize donations, tools, and volunteers. Over the course of a few weeks, in fact, she kicked organizational ass—putting together a barter currency network, helping to organize a living wage campaign, and spearheading a food-bank project.

Later, when Melissa and our activist friend Xubi Wilson decided to run for the County Commission on the Green Party ticket, we went all out. At the time, we had an eloquent and powerful Green Party platform in New Mexico (thanks to the party's Thomas Jefferson, my friend Steven J. Schmidt), and we'd already elected three registered Greens locally—two city councilors and Santa Fe's only municipal judge—but these had all been *nonpartisan* elections. County commissioners are expressly *partisan* seats in most states, including New Mexico. This is an especially important distinction in states where voting a "straight party" ticket is provided with the simple push of one button—as opposed to having to vote for each candidate as an individual.

We ran strong races, but that November my wife lost by a couple hundred votes in a two-candidate race and Xubi lost by more but did surprisingly well in his three-candidate race. More importantly, the election was widely covered in the press and ended up causing a substantial ruckus in local politics, so much so that within two years Melissa's top two platform pillars were passed into law. The first, New Mexico's new greywater law (see above), ended decades of bureaucratic hemming and hawing from the powers that be. The second, Santa Fe County's cistern ordinance, put the county on the frontier of water harvesting law in the United States.

Melissa and Xubi built a strong consensus around their new ideas and got an amazing amount of work done without ever having to take any of those mundane phone calls about hazardous potholes on county roads, noisy neighbors across the street, the latest subdivision graffiti spree, or any of those other important (but less than earth shattering) issues that constituents at the county level tend to have. Of course, past results are not proof of future performance, but my advice to anyone out there is to run for local office with water as one of your three or four key issues, and never lose sight of the fact that water harvesting can be great for the local economy, the environment, future generations, and the human soul. Either you will win and will be able to effect change from the inside, or you will lose and become free to make change from the outside.

To be fair to my father's perspective, it's worth mentioning that Melissa, Xubi, and I have a good friend named Fred Nathan. He's no outside agitator, but he plays a good, clean game of fostering change from within the system. In 1997, when he could have gone to work for his former boss, incoming Congressman Tom Udall (now New Mexico's junior senator), Nathan instead started a nonprofit organization called Think New Mexico. A results-oriented think tank that takes on one issue per year, Think New Mexico first decides what problem it wants to solve. Then the group develops a detailed and comprehensive solution to the problem. Next they start the slow and often painful process of pushing a bill through both houses of the legislature and finally to the governor for that often-critical signature. During its short history, the nonprofit has successfully pushed through a wide variety of laws ranging from the creation of full-day kindergarten in New Mexico to the repeal of the state's tax on food.

"The Strategic River Reserve is derived from a concept called the tragedy of the commons where everybody acts in their own self-interest," Nathan told me as we breezed through a maze of marble en route to a meeting of the Senate Finance Committee. "So we got together a diverse group of stakeholders from environmentalists to ranchers and got everyone to agree that it was in their own self-interest to get the state to lease some of the water in our most endangered rivers in order to prevent both ecological catastrophe and the loss of traditional forms of agriculture. So instead of tragedy," Nathan smiled proudly, "we've established a triumph of the commons."

Although it took two years instead of one to pass, Nathan's bill ultimately received 99 votes in favor of the Strategic River Reserve and nine votes against, which is about as close to consensus as one might ever wish for in a representative democracy.

Join an Organization

Not everyone can change careers; only a small minority of people want to become a cisterns-systems installer, and a scant few would ever run for public office. Even fewer would gear up as a water warrior to fight for water justice in a banana republic. Not everyone is cracked up to be a socially responsible investment advisor, a forward-thinking money lender, a new-urbanist architect, an environmentalist bureaucrat, a farm-school administrator, an enlightened high-school teacher, or a permaculture-design-course instructor. If you've also ruled out organic gardener, ecological landscaper, green builder, and alternative-septic-systems technician, don't despair! You can always do your part by joining an endless number of environmental organizations. Whether you volunteer for your local plant society or your church's ecology working group, or even if you simply send a few bucks to nature-walkers anonymous, it doesn't really matter. What's significant is that you participate in some way with a group that's out to do some good by protecting some river system somewhere, and what's really key is that you decide to increase your level of participation a little bit every year.

The goal is to give your chosen group some of your valuable time. It you don't have time, then give them a financial donation. If you don't have time or money, you can always talk loudly at social gatherings about the great work of your favorite groups. Technically, this type of blather should count for gradual-greening minutes, so if times are tough, get creative. Even if you merely help to attract what might seem like a small amount of attention to water issues in your bioregion, you will probably end up having a profoundly positive effect—far deeper than most people realize.

From an enjoyment and self-improvement perspective, the best kinds of groups are typically education focused. Your job is simply to decide what form of edification you want to receive and then start participating in your chosen group's tours, discussions, lectures, demonstration projects, classes, campaigns, newsletters, websites, or information booths. If there isn't something already available to you that you might enjoy, then invent something that's up your alley and get others to help you put it all together.

Many places have local botanical gardens, and these are great places to start. Like the Santa Fe Botanical Garden (SFBG), most are member-based organizations that survive in part with small annual contributions from people, your friends and neighbors, who care about local plant life. In addition to two nature preserves in the county, the SFBG has broken ground on a third site, an 11-acre parcel about a mile from

If there isn't something already available to you that you might enjoy, then invent something that's up your alley and get others to help you put it all together.

downtown Santa Fe. Leased from the City of Santa Fe and the state of New Mexico for one dollar a year, three of the 11 acres will be planted with the intention of creating outdoor "rooms" focusing on a variety of themes. "A walk through the Santa Fe Botanical Garden will be reminiscent of a walk through the Museum of International Folk Art just up the road," Executive Director Linda Milbourn said on her way out of town on a blue-sky Friday afternoon. "But instead of folk art from around the world, we'll be focusing on plant life from the local bioregion." Another similarity that the garden will have with the four museums that it will neighbor is that it will be open to the public, and a small fee will be requested for nonmembers. "As a means of saying thank you to our membership, walks through the garden will be free for anyone who has kept up with their dues," she told me as she drove along the ridge above Velarde.

With the help of Earthworks Institute, a local ecological nonprofit, about a half mile of the Arroyo de los Pinos has already been given a new lease on life. Check dams have been installed, mulch has been spread, and appropriate grading techniques have been implemented along a system of nature trails throughout the less-cultivated eight acres of the garden.

Botanical garden designer Gary Smith has also completed a master plan for the "rooms" and their associated "plant collections." Milbourn says "This part of the property will show people how to have stunningly beautiful gardens that use water responsibly. One aspect of the project involves harvesting roof water from the state's Stewart M. Udall Building, just up the hill from the garden's property line. Another exciting development," Milbourn continued on cue, "is that three adjacent neighboring properties have agreed to donate their blackwater to our orchard of heirloom fruit trees."

Milbourn said there are at least a dozen ways to get involved with the Santa Fe Botanical Garden and that this is the case with probably every botanical garden in the world: "We need docents to lead nature walks, volunteers to work with kids in our wetlands on the south side of the county, salespeople to take the reigns at our plant-sale booths, and we are always looking for experts in the field who can teach our members some of their skills."

Our local botanical garden also takes periodic field trips

to places of interest. In September of 2008, the SFBG organized a 30-person tour through a property that my company, Santa Fe Permaculture, had been working on for over a decade. Complete with swales, gabions, check dams, seeding, mulching, native plants, an underground cistern, and a recent experiment with the lop-and-scatter technique, Milbourn said of the visit, "People drove away from that tour really jazzed. They learned a lot of very effective techniques that they'll try in their own backyards.

"You know, a friend of mine once said, 'When times get rocky, people buy dirt,' and one thing that I'm observing is a curtailing of big travel plans. People are looking for stuff to do that's close to home, and, sure, the whole food-security issue is a big part of this, but even more compelling at this point is the sense of community that people gain from participating in our programs. You make a lot of new friends with a shared interest, and you get the kind of true companionship of like-minded people that we all deeply desire."

See? All that you really have to do to enjoy your time on this planet as a water harvester is to make sure you find an organization that provides useful skills and meaningful information. If you do, your newfound community of water- and environmentally conscious individuals will become your valuable ticket to the happiness and peace of mind that gradual greening promises.

Do the Math

It's easy for me to say, "Do the math"—especially when I tend to avoid math in the same way that I'd scamper from cat pee on a loveseat. Still, it's an important job that somebody has to do. Fortunately, when it comes to local water supplies, people like Consuelo Bokum gravitate toward doing the math for the rest of us. As the director of the New Mexico Water Project, Bokum and her many colleagues from numerous government agencies and nonprofit organizations work on an ever-evolving water plan for the local watershed.

"The goal of water planning," Bokum said at one of River Source's annual forums, "is first to determine if there is a gap between the available water supply and the current and future demand for water. If a gap exists, the next step is to come up with ways to shrink and ultimately eliminate that gap." Not surprisingly, Santa Fe's water plan showed a significant "water gap" throughout the region, "so we came up with five proposed solutions ranging from buying more water rights to limiting future development," she went on. "None of our proposed solutions was more cost-effective than water conservation. Right after the report came out, the people of this already water-conscious part of the world began to take conservation to a higher level, and one of the most innovative strategies to take hold has been water harvesting."

But the gap has by no means been eliminated, Bokum emphasized. "People used to talk about an aquifer the size of Lake Erie that supposedly sat under Albuquerque, but what seems to have happened is that historically people had drilled in some relatively productive well locations, and then the people who came later looked at the well data and assumed that this data would be the same pretty much everywhere in the Rio Grande basin. Unfortunately, current numbers are revealing an aquifer that's probably more than 20 times smaller than Erie."

Gradual greening suggests that with every passing year you add about 10 minutes of your daily free time to your commitment to water harvesting.

In addition to doing the math that focuses primarily on water quantities, in many places you will also want to do plenty of science in order to determine water quality. Your community might actually have Lake Erie under your feet, but if the source is polluted, it would be best to know this sooner rather than later. Of course, knowing and caring are two different things, so after doing the research please don't forget to do the outreach that will let your community know about your findings.

When news of our nation's pharmaceutically laced drinking water broke into the mainstream media back in 2007, my immediate thought was, *Gosh, maybe I'm wrong for believing in a steady and slow approach. Perhaps people will now want to move toward switching to cistern water sooner.* Now, since our society's reaction has been tepid at best, I have come to reaccept that the "gradual greening" approach still makes the most sense.

As you know by now, gradual greening suggests that with every passing year you add about 10 minutes of your daily free time to your commitment to water harvesting (or any other sustainable activity to which you feel drawn). In 30 years, you will be spending nearly four to five hours of your time per day saving the planet. If we were to add all of this free time to the green-jobs time that will be created during the current push to save the economy, then we are talking about the prospect of a culture that has achieved sustainability. But none of this will occur without realists like Bokum stepping up and sounding the charge on a regular basis. Not only did she and her colleagues come up with a well-documented report, they took their work to the people of the region and received loads of great press about it. "Without the outreach component, the project would have failed," Bokum said emphatically.

The hard part, of course, will be to incentivize so many donations of human free time. Looking on the bright side, even though the "drug cocktails in our water supplies" stories and local water plans won't cause the required social motivation on their own, this kind of research and media coverage will certainly help.

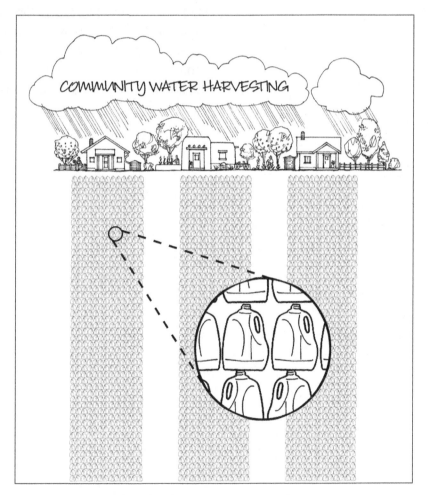

Neighbor Power

In contrast to the poisoned waters in our midst, keep in mind that precipitation is the cleanest type of water available, so, whenever practical, property owners ought to be encouraged to harvest it—especially in an age during which animal and human medications (not just our traditional wastes) are polluting our conventional water sources. Certainly, bacteria, viruses, and other nastiness can live on roofs and in tanks filled with water, but with some simple on-site filtration and a little regular monitoring, precipitation is the healthiest thing you can drink—with or without a prescription. As the water harvesting industry grows, we will need to bring modern science on board. In a democracy, this means getting active and in this case ensuring that research dollars get used for appropriate and beneficial projects.

Imagine if we were to spend 10 percent of NASA's budget on water harvesting. Sure, we might find a little less water on Mars, but if we increase this amount of funding by 10 percent every year, we would create enough green jobs and enough water resources to renew our nation's leadership role in the never-ending human quest to do what's right.

6

Awaken the Spirit of Change

Better than any argument is to rise at dawn
and pick dew-wet red berries in a cup.
—Wendell Berry, "A Standing Ground"

One thing I haven't done in this book is belabor you with tables, stats, charts, graphs (okay, one graph), and other forms of formulaic proof. Although I actually believe that water harvesting is the fountainhead of sustainability, I'm the kind of person who would just as soon avoid multiple citations, arrows, footnotes, and algorithms. Hey, even if I could dream up a mathematical proof of my theory, how would you know that my facts were accurate, my postulates were unprejudiced, my logic was sound, and my analysis was all encompassing?

Instead, I assumed from the beginning that you, dear reader, had an intuitive sense that water harvesting is something on which we, as a society, need to focus. In the same way that we need to teach young people to be polite, read, write, add, subtract, multiply, and divide, we also need to teach them about that which sustains us, that which we need every day for our survival—especially in our modern world, a world that wants and needs to be less estranged from nature.

On any journey, grown-ups set examples for children. We stop at red lights. We look both ways, and we show respect for and give adequate space to other drivers, cyclists, and pedestrians who might surprise us with their actions. As we start heading down the road toward this abstraction that we call "sustainability," what is the example that we should set?

Start to think about precipitation and water use in a new way. You are the change agent.

Spreading information about the importance of water and water harvesting is key, but it's not enough. Many of us also need to focus on homegrown and regionally produced food, alternative transportation, distributed power generation, eco-fabrics, green architecture, local economics, grassroots democracy, biodiversity, media diversity, and the rest. The list of things to do is endless, and it is this endlessness that often overwhelms us.

This is why I think the whole endeavor toward a greener planet has to start with something as specific and basic as you and your water. How else would we ever be able to jumpstart a shift in consciousness except through something that we all need regularly and something that we can all easily manipulate? With so much else to do, keep in mind that I am not asking you to change your life drastically overnight. I am asking you to start to think about precipitation and water use in a new way.

As long as our economic fortunes don't take too serious a tumble, and as long as atmospheric deterioration doesn't take too big of a toll, we have time to get our water harvesting act together. Just imagine how much we will be doing in 30 years if, on average, Americans spent as much time in their edible gardens, supported by harvested water, as they do gambling and playing video games. Sustainability would be much less of an abstraction, and in some cases it would be on its way to reality.

You are the change agent—committed to the cause by budgeting time every day, week, month, season, or year, and prioritizing how to employ your time best, how to save or how to keep busy with new ways of water harvesting that work well with whatever situation in which you find yourself. Where might a community focus its energy this year? What projects should you focus on the most this month? Which ones will we have to save for future years? What else can I do to harvest more rain—today?

We'll also have to admit to ourselves when we've slacked on our 10-minute gradual-greening commitments. How might we pay back time for these slipups? How might we bank time for a well-deserved vacation or sabbatical?

And we may always wonder if my proposed 10 minutes per day per year will be enough. If we discover that 20 or 30 minutes per day would have been better, then we'll adjust to make up for lost time. It's hard to pin down an exact time, too, because these decisions will be, by their nature, highly localized. I've chosen the 10-minute greening workup primarily because it's realistic and over time will produce measurable results in each of our lives and in aggregate for our communities and the larger, inextricably linked environment.

I've avoided burdening you with sad stories of multinational corporate malfeasance. I haven't dwelt on our planet's lack of water being a cause of world hunger. I have only mentioned the potential effects of global warming because it's such a hot topic these days and a well-known subject matter helps infuse ideas. But my goal throughout *Harvest* is to provide incentives to encourage people to take all of this water harvesting information to heart. I have kept the "gloom factor" down throughout, but now since we must part, I feel that it's my duty to sock it to you. Sorry.

The best argument for working toward sustainability comes down to this: If we don't grow gradually green, what are we going to eat?

I think of my Grandpa's infamous "hunk of horse" and I wonder how much of a joke it really was. He lived through the Great Depression—when a chunk of meat might have meant a much-needed source of protein for his family and friends. When times are extremely tough, how much will people care if dinner happens to be cow, squirrel, rabbit, or horse? Unless the horse has become the family's main source of transportation, my guess is not much.

My point is simply that these days a lot of signs are pointing to something worse than the Great Depression. Apart from all of our debt and staggeringly bad economic news, the biggest fear factor of all for me is that whenever the crash does arrive, in our current worldwide cultural state we no longer have any idea how to grow our own food. During the 1930s, if people were not still living on farms, they were relatively well connected to people who were. These days, there are just a few corporations controlling 95 percent of our food-production system.

If we don't start growing back at least a few of our agrarian roots soon, it's just a matter of time before we are forced to give up our freedoms in order to put the basics on the table. The edible landscapers and the backyard and market gardeners in each community, the "Reese Bakers," the "Mary Zemachs" and the "Patty Pantanos" of your community each play a role in forming self-sufficient, localized economies.

Get to know them and start to build a community around water and food issues in your bioregion. Whether you find an ideal group of like-minded people or not, the local-food (a.k.a. slow food) movement needs you to make these kinds of important life choices and become part of the solution.

Anyone can be an inventive landscaper, a dedicated permaculturalist, a determined eco-teacher, a backyard gardener, a seed saver, or a food-bank volunteer.

Anyone can be an inventive landscaper, a dedicated permaculturalist, a determined eco-teacher, a backyard gardener, a seed saver, or a food-bank volunteer. Teach your neighbors, friends, and children about cycles of life, of planting, nourishing, growing. Buy books by Alice Waters and Michael Pollan and take a tour of the best in food, which just so happens to be local. Get to know the people in your community who dedicate their lives to green work, local self-sufficiency, water problems, and health issues. They are the *influentials* of the movement.

One such person in my local community is farmer/ educator/radio personality Miguel Santistevan. Over the years, having had the pleasure of sitting on some of the same water-issue panels with Santistevan, I knew my interview would be fun, uplifting, serious, and extremely fast paced. I also knew that his words could help wrap up *Harvest the Rain* because no one I know understands water and community better than Miguel.

When I called him at the University of New Mexico, where he was earning his PhD, a pleasant woman's voice on the other end answered with one simple word, "Sustainability." I'd never heard anyone pick up the phone and do that, and the amazing part was that I wasn't surprised by this beautiful one-word greeting. It spoke of my friend's life work and spiritual calling, his vocation and his avocation.

The word *sustainability* is so much a part of the modern vernacular that it even sometimes makes it to the nightly news— sandwiched between crimes/celebrities and weather/sports. But this is the shocker: Right now, at this moment, we actually have a chance to act, each of us, to sustain the survival of our culture. We might not get another opportunity like this, and it's thanks to the self-described seed savers like Miguel Santistevan, who live, breathe, and create sustainability, that our chances go up.

"When global warming sets in," he said in his thoughtful and determined northern New Mexican staccato, "the rest of the world is going to resemble the brown *mesas* and hard soil we have here, so it'll be up to us to teach a heck of a lot of people how to survive. The problem is we've mostly forgotten how it's done. People thrived here for centuries before herbicides, pesticides, fungicides, and chemical fertilizers. There were no draft animals doing any of the work, no plows. They didn't even have metal tools. But the traditional farmer could look up at the mountains in the springtime and have a pretty good idea as to which crops were going to do best that summer.

"During the drought of 2002, I got two good waterings in before the ditch dried up completely. First, I watched my corn, then my squash, and then my beans go. Everybody just shriveled up, so I spent the summer focusing on my teaching and research, but I never bothered to check on my crops. The interesting thing is that when it came time to prepare my fields for winter, I went out there and I was amazed at how much edible plant material had survived. I had all kinds of things to eat.

"I had been working with native and drought-tolerant edibles for years, but that experience really woke me up. These days, I don't use as much water as I could because I would rather my crop provide a little less yield, if it is going to grow stronger and become better adapted for the following year.

"You know, everybody is running around worried about the future and wondering what could possibly be the fate of the human race. But I learned something long ago from a respected elder. 'Don't overanalyze the problem,' he said. 'The solution is simple: Gather your people and plant the fields.'"

Let's all decide *now* to make the necessary shifts in consciousness toward a greener Earth. Enjoy the process, learn as we go, and get better as time goes by. Produce water from the sky, be aware, be sustainable. Let's make real change, not progress for its own sake but a real reformation of ourselves, for ourselves, for future generations of people, and for every struggling species. Now you know how to start. All it takes is a little time, a small patch of land, a reasonably patient perspective, and a simple shift in our perception toward the sky.

Afterword

Just north and slightly west of the centuries-old farms along the Rio Grande, where this book began, a red, rocky *arroyo* rises to a *mesa* top high above a set of antediluvian mineral springs. Tewa legend has it that an epidemic sent the people packing from an ancient village there, where generations could go and soak in the hot water in between hunting expeditions and stints tending the extensive edible gardens that bordered their tall, thick, *pueblo* walls.

About 2,500 rooms once made up the Posi-ouinge, or "Village of Greenness." Now, all that remains are long mounds of earth, covered with cacti, saltbush, and sage. It's a strange place to walk because the ground around you is far from natural. It swells up in strange ways, and rises at your feet—at first too quickly and then not soon enough. Walking up the sides of the long-buried, long-abandoned earthen buildings and again back down into the 500-year-old *placitas*, it's easy for anyone brought up on *Planet of the Apes* to think of Charlton Heston's existential howl at the end of the original movie.

Upon finding the Statue of Liberty almost completely buried at the edge of a beach, the astronaut-protagonist finally realizes that he had not discovered a new planet (as he had previously believed) but, having spent centuries in space, he'd fallen back to an Earth where humans no longer ruled. I suppose one of the things that motivates me the most is that I don't want ever to feel responsible for the day when humans don't rule. As much as I love other animals, from fleas on the legs of insects to elephants and whales, I don't want to meet my maker with my tail between my legs knowing I could have done a whole lot better. If you feel this way too, this book has been written mainly for you.

During the writing of *Harvest the Rain*, I was often asked, "Who are you trying to reach? What is your target audience?" As best I can put it, *Harvest* is written for a rotating rainbow of readers with a wide variety of interests and concerns. In one part of the spectrum, you'll find construction-industry professionals looking for new techniques and markets. In another you'll discover homeowners hoping to acquire a greener and more ecological thumb. Down at the opposite end of the spectrum, community activists in need of a winning campaign issue come to mind. I hope anyone wanting to answer President Obama's call to action is now more fully inspired to "begin again the work of remaking America" and the world.

Although I come from a progressive perspective, sometimes the entrepreneur in me reveals some serious, deep-seated libertarian credentials. Even though I attend religious services only irregularly, I certainly believe in the power of spiritual activity on a variety of levels. Even if I bash corporations around a bit, some of my best friends work for very big companies, and I love them and really do hope they stop losing their jobs. Plus, I've been the president of a corporation myself since 1997.

I want these ideas to reach everyone from children to great-grandparents, from ultra neoconservatives to extremely unionist liberals, from active churchgoers to anarchist atheists, from people determined to save every penny to those willing to invest a little extra cash for their own benefit and for that of our planet and her people. Plumbers and electricians, parents and teachers, businesspeople and nonprofit types, community organizers, and established politicos are all more than welcome to dive into water harvesting and the concept of gradual greening. It may even be that *Harvest* is for the free-love, food-sharing hippie *and* the gun-toting canned-food hoarder in all of us. Still, in the same breath I should make clear that this book is also for that unconscious (and sometimes reticent) naturalist embedded in our genetic code—who has been mostly missing in our culture since we moved to cities from forests and farms.

Harvest is a "how-to" book at its core. It's full of many pages of specific information, but it does not provide many instructions that you *must* apply in order to prevent failure. Although I may not always convey it, I should be clear that I tend to approach my work with great humility, and rarely do I presume that my answer is the only possible one.

Mostly, this is a book for those who want to save the world, answer the call of history, or who simply want a better life, a healthier home, and a more productive community. The common denominator among these people may be that they have few options for acting toward attaining these goals. Water is a great context for many things, but as the lifeblood of consciousness, it holds a special power that seems to me to be revolutionary because it could easily become the material cause for the new progress that we now need to make.

It is time to change. Fortunately, there are people all over the globe ready to step forward. Our success will depend on our ability to quickly and continuously connect with each other. To this end, please do not hesitate to cast a line into the ever-changing stream at this book's website: www. harvesttherain.com.

Acknowledgments

You know in spite of all you've gained,
You still have to stand out in the pouring rain.
—Robert Hunter, "Sugaree"

nfinite thanks to Steven J. Schmidt for his watchful insight and consistent creativity, Barbara Doern Drew for her merciless editing and thoughtful input, Richard Lederer and Tom Knoblauch for their cleanup work and pun-loving fun, George Lawrence for his fastidious illustrations and loyal whimsicality, and Carl Condit and Jim Smith at Sunstone Press for their nearly constant trust and initial invitation. To my teachers (in order of appearance in my life), Anne Downey, Ken Carpenter, Paul Pilcher, Danny Kerr, Barnes Boffey, Ken Archer, Artemis Stamatopoulos, Timothy R. Burroughs, C. Brett Boocock, Jerome Garcia, Anne Downes-Catterson, Thomas R. Barrett, Richard Lederer, Clifford J. Gillespie, Jack Pulaski, Gary W. Hart, Sue Casey, Mark Steitz, Ralph Swentzell, David Starr, Charles Bell, Cary Stickney, Stephen van Luchen, Chester Burke, Elizabeth Engle, Jack Steadman, Bill Mollison, Derk Loeks, Roberto Mondragon, Ronald Gasca, Greg Nussbaum, Patricia Pantano, and Larry Littlebird, *muchas gracias por todos dichos*. To the many wonderful, hardworking folks whom I had the pleasure of interviewing for this book and to everyone in the fields of water harvesting, landscape design, community development, small-scale agriculture, and water justice, please keep up your essential work! We need you. To Gramma, Gram, both Grampas, Mom, Dad, Melissa, Liam, and Keenan, thank you for being you—*Harvest the Rain* would never have been written without your inspiration. Finally, to everyone unfamiliar with the subject of water harvesting, thanks for reading. I hope this book has motivated you to get your feet and hands wet.

Index / Alphabetical

Index / Topics

Community Opportunities

Corporations, for Profit

Organizations, Not-for Profit

Passive Harvesting Basics

People

Periodicals

Toxins / Waste

Wastewater Harvesting Fundamentals

CPSIA information can be obtained at www.ICGtesting.com
Printed in the USA
LVOW03s1835280115

424690LV00002B/29/P